This bo

RURAL ECONOMIC MODELLING

Rural Economic Modelling

An Input-Output Approach

Edited by

P. Midmore

Welsh Institute of Rural Studies
University of Wales
Aberystwyth, UK

and

L. Harrison-Mayfield

Department of Agricultural Economics and Management
University of Reading, UK

CAB INTERNATIONAL

CAB INTERNATIONAL
Wallingford
Oxon OX10 8DE
UK

Tel: +44 (0)1491 832111
Fax: +44 (0)1491 833508
E-mail: cabi@cabi.org
Telex: 847964 (COMAGG G)

A catalogue record for this book is available from the British Library.

ISBN 0 85199 112 2

Printed and bound in the UK by Biddles Ltd, Guildford

Contents

Contributors

ALASTAIR BAILEY: *Research Associate, Department of Agricultural Economics, Wye College, University of London, Ashford, Kent TN25 5AH, UK.*

LUCY HARRISON-MAYFIELD: *Research Associate, Department of Agricultural Economics and Management, University of Reading, 4, Earley Gate, Whiteknights Road, PO Box 237, Reading RG6 2AR, UK.*

SCOTT MCDONALD: *Lecturer in Economics, Department of Economics, Sheffield University, 9, Mappin Street, Sheffield S1 4DT, UK.*

PETER MIDMORE: *Lecturer in Agricultural Economics, Welsh Institute of Rural Studies,The University of Wales, Penglais, Aberystwyth, Dyfed SY23 3DD, UK.*

MORAG MITCHELL: *Lecturer in Economics, Economics, Marketing and Management Department, Scottish Agricultural College, Auchincruive, Ayr KA6 5HW, UK.*

DEBBIE ROBERTS: *Lecturer in Agricultural Economics, School of Economic Studies, University of Manchester, Manchester M13 9PL, UK.*

NOEL RUSSELL: *Lecturer in Agricultural Economics, School of Economic Studies, University of Manchester, Manchester M13 9PL, UK.*

DEMETRIOS PSALTOPOULOS: *Research Associate, Department of Agricultural Economics, The Agricultural University of Athens, 75 Iera Odos, Athens, 11855 Greece.*

KENNETH J. THOMSON: *Professor of Rural Economics, Department of Agriculture, Aberdeen University, 581, King Street, Aberdeen AB9 1UD, UK.*

Preface

This book is the product of a seminar held in Reading in 1994; most of the content consists of revised versions of the papers presented there, although the introductory and concluding chapters are completely new. We are extremely grateful for the co-operation of the contributors, and to George Peters for providing an introduction and also for chairing the original seminar. However, the overall responsibility for the form of this volume rests with us.

Peter Midmore
Lucy Harrison-Mayfield

Foreword

G.H. Peters
Research Professor in Agricultural Economics, The University of Oxford

For many years it has been understood that agriculture - narrowly defined as the farm sector engaged in the direct production of food and other products - is not an activity isolated from the remainder of the economy. In the USA in the mid-1960s, for example, it has been suggested that direct employment in farming of 5.6 million people was supported by 'backward-linking' supply industries employing about the same number (5.7 million), with the result that the production of raw material provided 'forward-linking' employment in processing, marketing and distribution for 12 million persons. The aggregate of 23.3 million made up no less than 30.4 per cent of the total workforce of 76.6 million persons. In France, in 1981, production agriculture directly involved around 8 per cent of the workforce, but the whole agriculture and food complex accounted for some 22 per cent of the country's labour force (Revel and Riboud, 1981, Chapter 5).

It is possible to contest such estimates and to seek to refine them, though it is obvious that there are important potential relationships between closely related sectors, and that events which affect one can have spillover consequences for others. The analytic methodology for the study of such relationships is input-output analysis, pioneered in the pre-war United States by Professor Wasily Leontief. In this book, so ably edited (and contributed to) by Peter Midmore and Lucy Harrison-Mayfield, the applications are to rural economic modelling, primarily in a British context, though having potential applications in any situation in which inter-industry or inter-regional relationships are being explored. The book is appearing at a time when interest in the British rural sector is becoming widely debated. It is noteworthy that *The Times* of February 9th, 1996 carried a letter, jointly written by The Prime Minister and the Leaders of the main opposition parties, advocating 'the protection of our countryside in its rich personality and character'. They did so at the instigation of the Council for the Protection of Rural England which, many years earlier in 1929, had solicited a similar letter from their forbears Stanley Baldwin, Ramsay MacDonald and David Lloyd George.

Despite the show of party unity the subsequent correspondence revealed wide differences of viewpoint about the nature of the policy responses which now appear appropriate from such correspondents as the urban planner, Professor J.B. Cullingworth, the environmentalist, Dr. Marion Shoard, and the former Liberal Party Leader, Sir David Steel. Some farmers entered the debate. One (Mr David Lort-Phillips) ventured to suggest that the party Leaders' attempt to 'spread a little harmony' had produced responses which 'degenerated into megaphone diplomacy', both from the environmentalist camp and from those who were more concerned with the problems of the farm sector, which revealed nothing more than an 'abyss of misunderstanding about countryside issues'. Another (Mr. Keith McDougall) suggested that the sheep industry alone (with all its direct and indirect employment) is responsible for 250,000 jobs, and that the addition of cattle and dairying could take the total to 1 million. Rural livelihoods, in short are important in themselves and are linked to others, and the Countryside Movement (a recently established group) is far from being representative of a 'reactionary rural minority'. Readers of this book, it must be stressed, will not be treated to a display of polemical invective. The tone, instead, is that of a careful academic treatise. It is worth remembering, nevertheless, that rural matters can arouse passionate debate and that it is all too easy to root arguments in prejudice rather than in careful analysis. The contribution, here, is to the latter.

A description of the contents of the volume can be found at the end of the editors' survey of the techniques (Chapter 1, pages 12-13). It is also noted there that the book stems from a seminar held at the University of Reading in 1994. The blizzard conditions which were experienced on that occasion did not deter participants from struggling to reach the venue or dampen their enthusiasm in tackling the intricacies of the subject. These are considerable, beginning at one level with the high mathematical content of the input-output approach, continuing through to general problems of data collection and presentation which is never easy even at the national level of application, and moving on to the notable difficulties of obtaining the factual information acquired for detailed application to agriculture, forestry and the rural economy.

Leontief is mentioned - the particular example of his work relates to pollution abatement which occurs in Peter Midmore's look at future directions for research - and it is worth recording that he once commended agricultural economists for having demonstrated 'the effectiveness of a systematic combination of theoretical approach with detailed factual analysis' (Leontief, 1971). Those attributes, which are well demonstrated, will be entirely evident in the forthcoming pages.

References

Revel, A. and Riboud, C. (1981). *American Green Power*, Baltimore: Johns Hopkins University Press.

Leontief, W. (1971). Theoretical assumptions and nonobserved facts, *American Economic Review*, 61, 1-7.

Chapter one:
Rural Economic Modelling: Multi-Sectoral Approaches

Peter Midmore[1] and Lucy Harrison-Mayfield[2]

[1] *Welsh Institute of Rural Studies, The University of Wales, Aberystwyth*

[2] *Department of Agricultural Economics and Management, The University of Reading*

The changes that rural economy and society have experienced in recent years are wide-ranging and complex, and stem in large measure from changes in socio-economic norms and values. Though reforms of agricultural policy have self-evidently played a part, other obvious influences have been the heightened awareness and revised political thinking on the environment, increased international competitive pressures and compensatory changes in the structure of social and regional funding.

These extensive interdependencies have been recognised, and there have been increasingly ambitious attempts to understand and predict their causal structure, to more precisely ascertain the future impact of planned or proposed policy changes. A recent, sophisticated attempt to represent many elements of social, economic and environmental processes in the countryside has been carried out by the Centre for Agricultural Strategy (Jones *et al.*, 1995), whose LUAM (land use allocation model for England and Wales) encompasses a number of different economic techniques, as well as being explicitly linked to natural resource conditions that have been determined from a large-scale environmental survey. Though this model has been under development since 1986, its essential character has remained the same: it is based on the allocation of 1 kilometre grid squares, which have been categorised according to the ITE land classification system in order to determine their agricultural potential. The model then allocates land of various types to production activities on the basis of net margins, by means of linear programming techniques. This core of the model can be solved either to optimise on the basis of contemporary policy variables such as prices or physical limits on production, or on the basis of future, conceivable policy states. From the results of this, other analyses can be bolted on, such as estimates of the indirect effects of changes in levels of agricultural output, consequences for the natural environment, and regional and sub-regional impacts.

The quality of results derived from this model are mixed. Although, on aggregate, predictions of agricultural output based on *ex ante* conditions do not diverge too widely from actual outcomes, the regional predictions are not as good. From a critical perspective, it is perhaps surprising that the model even does as well as this. It involves a number of stages, each with inherent tendencies to error and bias, and consequently each potentially compounding the inaccuracies on other stages on which they are dependent. This structure stems from the need to adequately represent the regional and sub-regional consequences of rural economic changes, especially on the natural environment. Given the level of resources available, it is difficult to model these separately and still arrive at a consistent aggregate result.

The provision of greater resources can result in substantially better performance at regional level. The NELUP model (aspects of which have been discussed by Allanson *et al.*, 1993; Moxey and White, 1994), which integrates extensive representation of hydrological and other processes within a water catchment area with economic and social considerations, effectively illustrates this point. However, in order to solve specific, pressing problems, models that can be rapidly and easily adapted to circumstances have proved much more cost-effective. Though traditionally input-output models have been seen to be costly, cumbersome and incompatible with the prevailing theoretical climate, a number of developments have made them considerably more appealing. This appeal is heightened through growing awareness of the importance of economic linkages, the social linkages that parallel them, and the dependence of rural regions on outside policies and markets. Within the input-output framework, these linkages can be extended beyond indirect effects to those affecting the environment, or even the inputs themselves (energy use or employment, for example). This point has been highlighted by Baumol and Wolff, who maintain that input-output analysis is important in rational policy formulation

> where two conditions hold: 1) inputs as well as output enter society's objective function directly . . . 2) where the economy in question is open, so that macroeconomic policy is largely powerless to increase employment to influence the use of other inputs.
>
> (Baumol and Wolff, 1994: 93)

Since it is obvious that, as far as agriculture is concerned, the indirect effects of changes as they impact on the rest of the rural economy are more important in many respects than the immediate or direct effects (consider, for example, the de-coupling of agricultural support in the recent CAP reform), it is not surprising that so much recent use has been made of the technique. However, Hamilton *et al.* (1991) have drawn attention to the caution that needs to be exercised in considering the measurement of these secondary effects, particularly the opportunity cost of the factors used, the effect of economies of size, the treatment of mobile factors and the inevitable short-term perspective of knock-on effects.

Although there may be difficulties in measuring such secondary effects, the importance of this technique stands out as the only methodology really capable of illustrating the whole economic system, and how changes in final demand in one industry filter through the system, affecting the other industrial sectors. This book provides an insight into the most recent applications of the technique to rural, natural resource based industries, and also to the countryside in general. As an introduction to the contributions that follow, this first chapter briefly describes the context from which these applications are drawn, building particularly on Midmore (1991a), who considered input-output modelling from a purely agricultural viewpoint. However, rural economics can be considered in two ways: first, as the economics of certain sectors, in a "sectoral" perspective; second, as the economics of rural regions, in a "spatial" perspective. We shall consider both perspectives, choosing sectors that have particular importance in rural areas, either from a land-use point of view (for example, agriculture or forestry) or from an ancillary point of view (the food sector), and examining models of small rural regions (for example, Wales, Grampian and Saskatchewan); these studies involve either exploration of the methodology or application of new methodologies to different situations.

Significant effort has been expended, under both headings, to overcome some of the standard problems of input-output analysis. These problems include the static nature of the input-output coefficients, the inclusion of induced (Type II) effects as well as the standard indirect (Type I) effects, the development of regional as opposed to national models, and the incorporation of price flexibility into modelling. As a result, our review is divided into four sections which aim to tackle each particular type of difficulty, before introducing the contributions that follow. These sections take on, respectively, the derivation of input-output coefficients, including the measurement and implications of technological change; Social Accounting Matrices, extended models and multiplier decompositions; spatial considerations dealt with by regional and inter-regional models; and models that deal with price and output effects simultaneously.

Input-output coefficients and structural change

Conventional input-output models use coefficients obtained from estimating the average quantity of input required to produce one unit of commodity or industry output. However, this average figure provides no estimate of the stability or variance of the coefficient, and it has been suggested that estimation by econometric means might provide a conception of their likely variability. One attempt to accomplish this, originally developed to overcome another difficulty – that of allocating input use to various enterprises from whole farm data – was undertaken by Errington (1987), based on earlier work by Tyler (1966). This involved the use of multiple regression techniques to estimate the proportion of total farm purchases used by each separate enterprise. In order to do this, each input was made the dependent variable for a multiple regression, with the various output levels as dependent variables:

$$b_i^t = \sum_{j=1}^{n} a_{ij} X_j^t + u^t \qquad [1.1]$$

where b_i^t is the total use of input *i* by enterprise *t*, and X_j^t is the output of commodity *j* by enterprise *t*. The resulting coefficients (the a_{ij}s) have, through their individual t-statistics, some indication of the reliability associated with them. However, this technique can also give rise to negative coefficients, probably caused by multi-collinearity. Panel methods are being developed for farm data which should ease this problem (for example see Bailey, Chapter seven in this volume); Bayesian approaches to this difficulty are also showing some promise (Moxey and Tiffin, 1994).

The stability of coefficients over time is of equal, if not greater importance than their reliability as estimates, especially where input-output models are used as forecasting tools. Coefficients vary as the technology of production changes, for example through investment in the modernisation of industrial plants or the substitution of novel inputs, which change the assumed linear equation of inputs required to produce one unit of output.

Some examples of models that seek to examine the effect of structural change within one or more sectors on the rest of the economy include Mules's work in Australia (1983); Lee's work in the United States (1991); and Barker's analysis in the UK (1991). These are all concerned with an examination of the role of agriculture in whole economy. More recently, Carri (1995) has completed a comparative analysis between different members of the European Union, and Lee and Schluter (1993) used the input-output framework to analyse growth and structural change of the food and fibre industries in the US economy 1972-1982.

Taking Lee and Schluter's work as representative, they built on methodologies that have been developed in Europe (by, for example, Forssell, 1988, 1989; and Skolka, 1989), apportioning changes in output to changes in (1) domestic final demand (2) export demand (3) inter-industry demand and (4) domestic supply ratios. Their disaggregation of the agricultural sector into 16 sectors is of particular importance in reducing bias in estimating the volume of outputs from the values and retail price indices and also in accurately identifying causes of changes in gross output. This difficulty was emphasised particularly by McDonald *et al.*' (1991) study of structural change and productivity growth in the agricultural and food chain 1954-1984. In this instance, the need for data classified by the same method over a long period of time exaggerated the difficulty. They used a Vertically Integrated Sector (VIS) framework as a method to estimate the relative efficiency of delivery of each category of final demand. Mattas and Shrestha (1989) have also examined interdependencies between industrial sectors, in this instance between food and non-food sectors, analysing the contribution of the food sector, disaggregated into raw and processed sectors, to economic growth.

Gould (1986a) examined the impact of structural change on resource use, using variables other than values of output, for example on the environment, energy use and pollution. Traditional input-output tables represent flows between industrial sectors in terms of the money value of output. Whilst it is possible to estimate the impact of changes by simply translating the estimates of output from each sector into, for example, energy use or employment by sector, Gould's analysis of the Saskatchewan economy actually measured the flows of the other resources themselves, modelling flows of energy between sectors. Hence changes influenced by final demand could be separated from those influenced by technological change, and further, the effect of technological change could be subdivided into changes in market share and in input structure.

The problem of secondary products in the use of input-output tables has been discussed at length in the literature: industries produce more than one product, yet the input-output matrix only has room for a single set of technical coefficients for each. The standard method to accommodate this fact (see Stone *et al.*, 1962, for the original, comprehensive description) is to construct Use and Make matrices, and to combine them according to either industry or commodity technology assumptions, or a mixture of both:

- the industry technology assumption is that commodities are produced using the technology of the producing industry, as opposed to the industry that produces the commodity as its principal product;
- the commodity technology assumption is that the commodity produced uses the technology of the industry that produces it as its principal product.

Agricultural studies tend to use the industry technology assumption as it is more applicable to farm systems. However, it is possible to create 'hybrid' synthetic industry by industry and commodity by commodity matrices (Roberts, 1992). Roberts argued that the former is preferred for employment, income distribution and sectoral planning analyses, whereas the latter should be preferred for investigating international trade and linkage analysis. The commodity technology assumption versus the industry technology assumption is also discussed in a recent paper by Lee (1991) which undertook a critical analysis of the difficulties in using the United Nations Systems of National Accounts (SNA) procedures for construction of input-output tables, with reference to US agriculture.

Social accounting matrices and decomposition of multipliers

The standard input-output model takes into account only Type I effects, which are the indirect impact of changes in final demand from production industries on other production industries. Changes in household demand are aggregated with the rest of final demand and so are exogenous to the system. However, these changes – for example in the agricultural sector – will affect the household income of farmers, farm workers and others, and therefore household demand for other goods, a proportion of which will be purchased in the local economy and will consequently have a large knock-on effect through the system. It is possible to incorporate the households as an endogenous part of the system and so to measure not only Type I effects but also Type II effects (direct, indirect and induced effects) thus creating a less open input-output model. The mechanics of this extension are described, for example, by Pyatt and Roe (1977), Pyatt and Round (1977), Bulmer-Thomas (1982) and Roberts (1991). By incorporating households into the Use matrix, it is then possible to disaggregate it into different groups (by income, or other socio-economic stratification) so long as information on spending and purchasing patterns for each group can be established. Such a matrix, more formally a Social Accounting Matrix or SAM, can be used to model the effect of changes in exogenous final demand on the purchasing power of different types of household, and how this feeds back through the economic system.

SAMs have been used extensively in developing countries, particularly to analyse the distributive effects of policies between rural and urban households (Pyatt and Round, 1977). Cohen (1989) examined the use of the SAM framework for gaining insight into development policy on issues of growth and distribution by comparing the multipliers from ten countries. He found, for all ten countries studied, the agricultural growth multipliers show more progressive redistributive effects than other manufacturing industries.

They have also been used to analyse the importance of an industrial sector to the economy. Research by Adelman and Robinson (1986) examines the contribution of the agricultural sector to the United States economy using a SAM framework. While recognising the contribution this model can make to exploring the important structural features of US agriculture, and providing upper bounds on the quantitative impact of various types of interventions intended to benefit farmers, they recognise the benefit of using this as a basis for constructing a computable general equilibrium model (CGE), which would incorporate price responsiveness and demand behaviour.

Batey *et al.* (1987) examine how the results of economic impact analysis are likely to be influenced by the choice of household disaggregation scheme. They use data from the Greater Cork region in the Irish Republic and examine impacts using four types of model. First they use the simplest form of model where households are treated exogenously (to produce multipliers showing direct and indirect effects only), then they follow the conventional response to inclusion of Type II multipliers – enlarging the input-output model by adding one row and one column to the matrix to represent household income and spending, endogenising them in the system of accounts (showing direct, indirect and what are termed intensive effects). However, this extends the assumption of linear and homogeneous consumption functions to each household, and also the same patterns of unemployment and migration. The third stage breaks down the latter assumption, dividing households between indigenous and immigrant populations (producing multipliers showing direct, indirect and extensive effects). The fourth, final stage of the model disaggregates once again to represent the redistributive elements of income change, explicitly and separately modelling the propensities to consume of the unemployed workers, including the indigenous unemployed (and producing multipliers that include direct, indirect, extensive and redistributive effects). They found that the largest component of the induced income effect was the result of extensive income, derived from immigrant households, though the intensive and redistributive income effects more or less cancelled each other out. However, for employment impact, there was a clear gradation of effects: extensive effects were the largest, but these were modified substantially by the redistributive effects. Round (1989), using examples mainly from poor countries, explored decomposition of input-output and economy-wide multipliers in a regional setting: this alternative method was demonstrated using flows between two regions and the 'rest of the world'.

Another limiting factor of input-output models, leading to overestimation of multipliers, is the assumption that multiplier effects are caused solely by changes in final demand. Multipliers are the estimated outcome resulting from a change of one unit of final demand on an industrial sector, as it feeds through the economy. However, changes may also originate in gross output, and by confining analysis to final demand changes means that these are ignored. This problem is tackled by Szymer (1992) where a 'total flow approach' is developed. This expresses the total effect of a change in gross output rather than final demand deliveries. Roberts (1994) advanced a version of the approach in an examination of the impact of milk quotas. She used a modified, SAM-based Leontief model to estimate the employment implications of milk quotas on the UK economy. Since milk quotas obviously act to change gross output in the dairy industry, rather than to change final demand, the modified model made output from the dairy industry exogenous to the system. This method eliminated overestimation of the knock-on effects of quotas, demonstrating a total full-time equivalent job loss of 11,000 (6,700 through backward linkages, 4,300 through forward linkages).

Spatial disaggregation: regional and inter-regional models

Using input-output tables to explore regional links

The models described up to this point have, for the most part, relied on estimates of input-output relationships for countries as a whole, though industries may show considerable spatial variation in their input-output patterns. The consequences of certain economic events for specific rural regions or zones are important for rural economic models, and input-output tables have been created for smaller areas in order to facilitate their construction, and enable the effects of changes in the economy to be examined at a more local level. The analyst is faced with an invidious choice, either to collect regional data to compile the tables, or to adjust national (or supra-regional) tables by mechanical methods. Usually, a compromise is sought between the two methods, as neither are completely satisfactory. Compiling tables from survey data requires huge quantities of data that are often difficult, or sometimes even impossible to obtain. Several modification techniques have been devised to adjust the national input-output coefficients to those that more aptly describe the region (see Richardson, 1972, or Reifler, 1973 for a review). However, none of the methods can provide an accurate enough representation of the actual relationships to be of real value. Therefore, the path frequently taken in creating a regional table is first to adjust the table mechanically, then weight the coefficients with the survey data that are available.

This has been a fruitful field for research, and innovations in the construction of regional input-output tables range from Moore and Petersen's (1955) prototypical work in the United States, to that of Gould and Kulshreshtha (1986) in Canada. In Wales, Midmore (1988) developed a regional input-output model that disaggregated the agricultural sector and allied agricultural industries. In addition to the standard multiplier analysis he used the model to show the income effect of a 1 per cent price rise of various commodities. Gilchrist and St.Louis (1994) explored the effect of diversification policies on the small region of Saskatchewan, assuming that the diversification would improve the performance of the economy by making it less sensitive to outside shocks. The implications of policy alternatives were tested by examining the mean and variance of regional income to obtain partial policy rankings on the assumption that the criterion is locally increasing in expected regional income and decreasing in the variance of regional income.

There is a considerable tradition of work on regional input-output modelling at the Aberdeen School of Agriculture. For example, Johns and Leat (1987) created an input-output model of the Grampian region in Scotland using the GRIT (Generation of Regional Input-Output Tables) technique, pioneered by Jensen *et al.* (1979), to estimate regional input-output coefficients for the Grampian region of Scotland. Their technique combined use of Simple Location Quotients (which are a measure of the relative importance of each industry regionally compared with its national importance) and Cross Industry Location Quotients (a measure of the relative sizes of selling and purchasing industries that determines whether the selling industry will be able to provide enough goods to the purchasing industry in that region) with survey data. Output, income and employment multipliers, both Type I and Type II, were calculated for 34 industrial sectors. The results showed that although the largest initial impact on incomes and employment of increasing output in any one sector is in the service sector, and the lowest impact in the food sectors, this was reversed once the indirect and induced effects are examined. Selected output and employment multipliers are shown in Table 1.1.

These results highlight the importance of examining the relative size of the direct and indirect coefficients, as well as their ratio, a point discussed by Mattas and Shrestha (1991). They suggest a new approach – that of input-output elasticities – which incorporates both multiplier effects and the relative size of economic sectors.

The original Grampian study was followed up by Leat and Chalmers (1991), in which the agricultural sector was disaggregated into eight separate sectors by using multiple regression procedures on farm survey data (see Errington, 1987; Midmore, 1991b). They concluded that with every job lost on the farm, approximately one off-farm job will be placed at risk. The largest output multipliers are found in the pig, poultry and slaughtering sectors. These sectors also have large employment multipliers, along with cattle and milk processing. They concluded that the assumption of constant input-output relationships biased results when estimating large-scale changes in agricultural and food production, and also that the model functioned better to predict employment change for contractions of industry than expansion. This was ascribed to the readiness of business to shed labour in times of difficulties but a certain reluctance to take on new staff in times of expansion.

Table 1.1. Income and employment coefficients and multipliers of selected industries in Grampian.

Industry	*Direct, Indirect & Induced Coefficient*	*Direct Coefficient*	*Type II Multiplier*
Income			
Forestry	1.157	0.719	1.61
Public Administration & Other Services	1.108	0.795	1.39
Financial and Business Services	0.986	0.636	1.55
Retail Distribution	0.740	0.425	1.74
Motor Trades	0.729	0.398	1.83
Mechanical Engineering	0.658	0.379	1.72
Employment			
Public Administration & Other Services	227	164	1.38
Forestry	204	120	1.70
Retail Distribution	193	135	1.43
Hotels & Catering	179	123	1.45
Ore Extraction	146	91	1.61
Oil & Gas Extraction	146	107	1.36

Source: Johns and Leat (1987: 254).

Because of the traditional predominance of the agricultural industry in rural areas, inevitably most of the research into the workings of the rural economy has concentrated on this particular sector. However, other sectors have also featured in such studies, particularly forestry. Psaltopoulos and Thomson (1993) were responsible for a recent example of this type of research, examining the role of forestry in rural development in Scotland. This employed the GRIT technique to construct input-output tables for rural Scotland, in order to estimate backward linkages for the forestry industry.

Using input-output tables to explore links between regions

Inter-regional models examine the flows of goods and services between different regions as well as to the 'rest of the world'. A good description of their compilation appears in Richardson (1972), although Gould's (1986b) study of the impacts of Prairie Branch Line Rehabilitation is more recent. This represents the Canadian economy as four regions – Manitoba, Alberta, Saskatchewan, and 'Other Canada' – and data were collected to represent not only the transactions between these regions, but also the level of international and inter-regional trade for each commodity. Such a model allows the total impact of a project to be estimated, and in this instance, used as the basis of a complete cost benefit analysis study. Similarly, Bernat and Johnson (1991) used input-output models with disaggregated household sectors to show large differences between household linkages in small rural regions in Virginia, USA.

Since input-output models are conventionally demand-driven, for the most part analysis concentrates on backward linkages, and might therefore neglect elements of the total multiplier effect. It is possible to examine forward linkages, as demonstrated by Henry and Schluter (1985), and Harthoorn and Wossink (1987), though the methodology is less straightforward. Roberts, in a recent analysis on the effects of milk quotas on the UK economy (1994), introduces the concept of forward linkages into her SAM of the UK. Her approach was based on Jones (1976), who provided a method of calculating output coefficients, which can then be inverted in much the same way as the input coefficients, to determine the forward multipliers.

At Reading, ongoing work to determine the employment implications of agricultural change in the UK at a regional level uses a social accounting framework where households are decomposed to the five income groups readily available in the Food Expenditure and General Household surveys. This builds on and updates the work of Roberts, using the new 1990 input-output tables which became available for use in June this year.

Input-output and econometric modelling

The relationship between neoclassical and input-output economics is somewhat insubstantial. Neoclassical economics is concerned with value and how this can be related to the endowments and technology of an economy, making considerable use of the concept of marginal productivity. Input-output economics examines the transmission of effects from change in one part of the economy through the concept of the technical coefficient, approximating in a rough and ready but workable form of a Walrasian general equilibrium model. Hence, input-output economists are criticised for the rigidity of the input-output model, particularly the fixed nature of the coefficients and failure of explain factor rewards, while, on the other hand, neoclassicists have no alternative framework through which to analyse effects on the whole economy. It is possible, though, to move toward a synthesis of the two approaches by using the neoclassical economics to determine the price and value equations, then to use these within the input-output framework to examine the effects of shocks to the whole system (Raa, 1995).

One of the most limiting facets of the standard input-output model is that it only deals with changes in volume of output, and assumes prices to be fixed. Alternatively, its dual form can deal with the impact of price changes but only at the expense of an assumption of fixed output levels. Yet most changes in the real world, particularly as far as agriculture is concerned, deal with changes both in outputs and in prices. There have been attempts to incorporate prices into input-output models: Gigantes and Hoffman (1972), for example, used an iterative solution method for successive runs of price and quantity input-output models, though they acknowledged that there was no guarantee of convergence in this essentially *ad hoc* approach.

A more promising set of developments is offered through hybrid modelling. Hallam *et al.* (1994) have suggested a combination of traditional partial equilibrium analysis, within the framework of the input-output model, to ensure the incorporation of feedback effects. Diekhauer *et al.* (1979) established a method of combining econometric forecasting of final demands with estimates of linkage effect based on input-output multipliers; Bingley *et al.* (1985) have examined the impact of milk quotas on the UK dairy industry, in which the supply of milk is initially modelled using the conventional econometric approach and the results used to estimate employment and incomes in the dairy industry through the framework of the input-output approach; and Jones and Midmore (1990) have combined econometric modelling of final agricultural demand with estimation of regional feedback effects in a disaggregated input-output model emphasising agricultural sectors. The combination of the two approaches thus overcomes some of the disadvantages of the input-output modelling approach (the assumptions of fixed input structure, linear supply response, etc.) and incorporates its capacity to encompass the feedback effects through supply industries.

An approach that models both price and volume changes simultaneously is provided through the use of the more coherent Computable General Equilibrium (CGE) model. Haggblade *et al.* (1991) examine the inclusion of price in their article which compares alternative methods of estimating agricultural growth multipliers, and introduces their price endogenous model. This model allows upward sloping supply curves for non-tradables and for substitution between its inputs, rather than insisting on fixed coefficients. As a result of making prices endogenous in the model, they suggest that input-output methods might overstate the magnitude of growth multipliers by a factor of between 10 per cent and 25 per cent.

As far as CGE models are concerned, these are built on fundamental micro-economic principles and include non-linear feedback mechanisms. They tend to deal with highly aggregated industrial sectors (since their appetite for both computing power and memory is prodigious) and thus are not really suitable for individual sectoral analysis: rather, they are mostly associated with efficiency questions and neoclassical welfare analysis, for example analysing the effect of the GATT agreement (Kilkenny and Robinson, 1990) or in developing countries to determine the impact of alternative development strategies on the structure of production or distribution of income (Adelman and Robinson, 1986; Clarette and Roumasset, 1986; Howitt, 1986; Coxhead and Warr, 1991). The static nature of the models means they are not used to trace the effects of dynamic change after policy change. Their size and complexity means they have huge and detailed data requirements and thus are expensive to maintain and keep up-to-date, further reducing their flexibility.

Rural economic modelling: multi-sectoral approaches

The contributions to this volume all, in at least one respect, develop from the context set out in this chapter. All but one (the final, concluding chapter) were contributed to a seminar held in Reading in 1994; both this framework discussion and the epilogue to the book as a whole have drawn on discussions which took place at that time, though they have obviously also drawn on a wider range of ideas and sources. The linking themes to all have been derived from the multi-sectoral approach which was emphasised by the seminar and is the focus of this book. In this final section, we briefly indicate how each contribution links into the overall theme.

Lucy Harrison-Mayfield investigates the potential and practical difficulties associated with input-output modelling as a means of assessing agriculture's contribution to the rural economy, especially in terms of rural employment. This is in response to considerable uncertainty in agriculture itself, especially since the structure of support is undergoing vigorous debate and continual modifications. Her chapter describes the obstacles that need to be overcome to produce a rural-urban inter-regional input-output flow matrix. The major conclusions drawn are that, since difficulties of data availability preclude a spatial division of the economy into rural and urban input-output relations, and inter-regional flow modelling based on that division, an approach that addresses local problems in case studies, as they arise, might be more fruitful; an illustration of this principle might, for instance, be to build small-scale models of a rural town and the surrounding network of the farm economy.

Morag Mitchell provides an insight into how input-output models may be extended within the spatial context, using them in combination with a Geographical Information Systems approach to the investigation of pluriactivity. Selecting a specific set of non-traditional farm-based activities in tourism, the method she develops indicates how precise spatial targeting could assist the optimum location of activities from the point of view of achieving rural policy targets for economic activity and population levels.

Ken Thomson and Demetrios Psaltopoulos also address methodological issues, but in this instance in relation to forestry. Here, the long production cycle causes difficulties in allocation of input during different phases of forestry operations. Though the scales of activity in production and harvesting are clearly linked, this long-term relationship sits uneasily with the predominantly short- to medium-term focus of input-output modelling. In particular, technological changes will overtake forecasts very rapidly, but inter-industry effects might be considered crucial in the policy decisions that are based upon such time scales.

Debbie Roberts and Noel Russell open up the discussion by examining the relationship between income distribution and the structure of agricultural production in the light of disaggregation of the household sector, both in terms of consumption and labour supply. This develops the Social Accounting Matrix approach specifically to deal with the circumstances of agriculture in the United Kingdom.

Scott McDonald develops his and others' (1991) contribution to the earlier volume on this theme, to analyse structural change in the food and agricultural industries, accounting for changes in demand and improved efficiency in the use of factors of production. Restructuring of demand and technological changes can be demonstrated to have slowed the pace of agriculture and food industry productivity improvement in the early 1980s, though submerged beneath these results are potentially more interesting phenomena, related to product quality and differentiation which cannot be demonstrated until more disaggregated data are made available.

Alastair Bailey revisits a theme that has been of considerable interest to linear programmers and farm business analysis as well as input-output analysis: the problem of deriving fixed input coefficients for activities that produce multiple outputs, and the implications for the accuracy of forecasts based on them. The alternative is set out as generalised production function approaches based on substitutability of inputs in response to demand changes, which would move the style of modelling closer to the CGE systems mentioned earlier.

Peter Midmore attempts to predict future focal points of interest for modellers in this tradition. Three major themes are identified: the development of suites of impact models for specific forecasting purposes, and the place of input-output, SAM and CGE models within them; the growing importance of environmental impact assessment and the possibility of integration of input-output approaches in bio-economic models; and the use of multi-sectoral modelling in highlighting the importance of social, cultural and economic sustainability alongside that of the natural environment.

References

Adelman, I. and Robinson, S. (1986). US agriculture in a general equilibrium framework: analysis with a social accounting matrix, *American Journal of Agricultural Economics*, 68, 1196-1207.

Allanson, P., Moxey, A. and White, B. (1993). Agricultural nitrate emission patterns in the Tyne catchment, *Journal of Environmental Management*, 38, 219-232.

Barker, T.S. (1991). Sources of structural change in the UK agriculture and food industries 1979-1984. In: Midmore, P. (ed.) *Input-output models in the agricultural sector*, Aldershot: Gower, 83-99.

Batey, P.W.J., Madden, M. and Weeks, M.J. (1987). Household income and expenditure in extended input-output models – a comparative theoretical and empirical analysis, *Journal of Regional Science*, 27, 341-356.

Baumol, W.J. and Wolff, E.N. (1994). A key role for input-output analysis in policy design, *Regional Science and Urban Economics*, 24, 93-114.

Bernat, A.G. and Johnson, T.G. (1991). Distributional effects of household linkages, *American Journal of Agricultural Economics*, 73, 326-333.

Bingley, P., Burton, M. and Strak, J. (1985). Inter- and intra-sectoral effects of milk quotas in the UK milk industry, *European Review of Agricultural Economics*, 12, 411-430.

Bulmer-Thomas, V. (1982). *Input-output analysis in developing countries: sources, methods and applications*, Chichester: Wiley.

Carri, C.B. (1995). Sources of output growth and agricultural performance in EU countries – 1960 to 1980, *European Review of Agricultural Economics*, 22, 231-248.

Clarete, R.L. and Roumasset, J.A. (1986). CGE models and development policy analysis: problems, pitfalls and challenges, *American Journal of Agricultural Economics*, 68, 1212-1216.

Cohen, S.I. (1989). Multiplier analyses in social accounting and input-output frameworks: evidence for several countries. In: Miller, R.E., Polenske, K.R. and Rose, A.Z. (eds.) *Frontiers of input-output analysis*, New York: Oxford University Press, 79-99.

Coxhead, I.A. and Warr, P.G. (1991). Technical change, land quality, and income-distribution: a general equilibrium-analysis, *American Journal of Agricultural Economics*, 73, 345-360.

Diekhauer, G., Schuman, J. and Meyer, U. (1979). Simultaneous determination of sectoral production quantities and prices in a dynamic input-output model – demonstrated for the Federal Republic of Germany, paper presented to the 7th International Conference on Input-Output Techniques, Innsbruck.

Errington, A.J. (1987). Labour use and labour requirements in UK agriculture, *Journal of Agricultural Economics*, 38, 271-279.

Forssell, O. (1988). A decomposition technique for analysing structural changes in production, *Discussion Papers on Structural Analysis of Economic Systems*, No. 8, Cambridge: Cambridge Growth Project (Department of Applied Economics, University of Cambridge).

Forssell, O. (1989). The input-output frame for analysing transmission of technical progress between industries, *Economic Systems Research*, 1, 429-446.

Gigantes, T. and Hoffman, R. (1972). A price-output nucleus for simulation modelling. In: Brody, A. and Carter, A.P. (eds.) *Input-output techniques*, Amsterdam: North-Holland.

Gilchrist, D.A. and St.Louis, L.V. (1994). An equilibrium-analysis of regional industrial diversification, *Regional Science and Urban Economics*, 24, 115-133.

Gould, B.W. (1986a). The impacts of structural change within an economy on resource use: an input-output analysis, *Applied Economics*, 18, 457-477.

Gould, B.W. (1986b). The impacts of Prairie branch line rehabilitation: an application of interregional input-output-analysis, *Canadian Journal of Agricultural Economics-Revue Canadienne d'Economie Rurale*, 34, 313-330.

Gould, B.W. and Kulshreshtha, S.N. (1986). An interindustry analysis of structural change and energy use linkages in the Saskatchewan economy, *Energy Economics*, 8, 186-196.

Haggblade, S., Hammer, J. and Hazel, P. (1991). Modelling agricultural growth multipliers, *American Journal of Agricultural Economics*, 73, 361-374.

Hallam, D., Midmore, P. and Peston, M. (1994). *The economic impact of the British beet sugar industry*, Reading: Department of Agricultural Economics and Management (University of Reading).

Hamilton, J.R., Whittlesey, N.K., Robison, M.H. and Ellis, J. (1991). Economic-impacts, value added, and benefits in regional project analysis, *American Journal of Agricultural Economics*, 73, 334-344.

Harthoorn, R. and Wossink, G.A. (1987). Backward and forward effects of Dutch agriculture, *European Review of Agricultural Economics*, 14, 306-334.

Henry, M. and Schluter, G. (1985). Measuring forward and backward linkages in the US food & fibre system, *Agricultural Economics Research*, 37, 33-39.

Howitt, R.E. (1986). Optimal control of general equilibrium models: discussion, *American Journal of Agricultural Economics*, 68, 1219-1221.

Jensen, R.C., Mandeville, T.D., and Karunaratne, N.D. (1979). *Regional economic planning: generation of regional input-output analysis*, London: Croom Helm.

Johns, P.M. and Leat, P.M.K. (1987). The application of modified GRIT procedures to rural development analysis in Grampian region, *Journal of Agricultural Economics*, 38, 243-256.

Jones, L.P. (1976). The measurement of Hirschmanian linkages, *Quarterly Journal of Economics*, 90, 323-333.

Jones, M.E. and Midmore, P. (1990). *An approach to forecasting the impact of changes in agricultural policy on employment and income in Wales*, unpublished report for the Welsh Office, Aberystwyth: Department of Economics and Agricultural Economics (University of Wales).

Jones, P.J., Rehman, T., Harvey, D.R., Tranter, R.B., Marsh, J.S., Bunce, R.G.H. and Howard, D.C. (1995). *Developing LUAM (Land Use Allocation Model) and modelling CAP reforms*, CAS Paper No 32, Reading: Centre for Agricultural Strategy (University of Reading).

Kilkenny, M. and Robinson, S. (1990). Computable general equilibrium-analysis of agricultural liberalization: factor mobility and macro closure, *Journal of Policy Modeling*, 12, 527-556.

Leat, P.M.K. and Chalmers, N. (1991). Analysing agricultural adjustment in Grampian using an input-output model of the agricultural and food complex. In: Midmore, P. (ed.) *Input-output models in the agricultural sector*, Aldershot: Gower, 57-71.

Lee, C. (1991). Recent development in construction of input-output tables with use and make matrices: an application to US agriculture, *Canadian Journal of Agricultural Economics*, 39, 795-803.

Lee, C. and Schluter, G. (1993). Growth and structural change in US food and fiber industries: an input-output perspective. *American Journal of Agricultural Economics*, 75, 666-673.

Mattas, K.A. and Shrestha, C.M. (1989). The food sector and economic growth, *Food Policy*, 14, 67-72.

Mattas, K.A. and Shrestha, C.M. (1991). A new approach to determining sectoral priorities in an economy: input-output elasticities, *Applied Economics*, 23, 247-254.

McDonald, J.R.S., Rayner, A.J. and Bates, J.M. (1991). Productivity growth and structural change in agriculture and the UK food chain: 1954-1984. In: Midmore, P. (ed.) *Input-output models in the agricultural sector*, Aldershot: Gower, 101-129.

Midmore, P. (1988). The impact of agricutural policy on income and employment in Wales: an input-output approach, unpublished PhD thesis, Department of Agricultural Economics and Marketing, University of Wales, Aberystwyth.

Midmore, P. (1991a). Input-output and agriculture: a review. In: Midmore, P. (ed.) *Input-output models in the agricultural sector*, Aldershot: Gower, 1-20.

Midmore, P. (1991b). *Problems in estimating agricultural input-output coefficients by regression methods*, Aberystwyth: Department of Economics and Agricultural Economics, Aberystwyth Rural Economics Research Papers, No 91-02 (University of Wales).

Moore, F.T. and Petersen, J.W. (1955). Regional analysis: an inter-industry model of Utah, *Review of Economics & Statistics*, 38, 368-383.

Moxey, A. and Tiffin, B. (1994). Estimating linear production coefficients from farm business survey data: a note, *Journal of Agricultural Economics*, 45, 381-385.

Moxey, A. and White, B. (1994). Efficient compliance with agricultural nitrate pollution standards, *Journal of Agricultural Economics*, 45, 27-37.

Mules, T.J. (1983). Input-output analysis in Australia: an agricultural perspective, *Review of Marketing and Agricultural Economics*, 51, 9-30.

Psaltopoulos, D. and Thomson, K.J. (1993). Input-output evaluation of rural development: a forestry-centred application, *Journal of Rural Studies*, 9, 351-358.

Pyatt, F.G. and Roe, A.N. (1977). *Social accounting for development planning with special reference to Sri Lanka*, Cambridge University Press: Cambridge.

Pyatt, F.G. and Round, J.I. (1977). Income distribution and input-output: an analysis for Sri Lanka, Warwick economic research papers ; 102, Department of Economics, University of Warwick.

Raa, T. (1995). *Linear analysis of competitive economies*, London: Harvester Wheatsheaf.

Reifler, R. (1973). Inter-regional input-output – a state of the art survey. In: Judge, G.G. and Takayama, T. (eds.) *Studies in economic planning over space and time*, Amsterdam: North Holland, 133-162.

Richardson, H.W. (1972). *Input-output and regional economics*, New York: Halsted Press.

Roberts, D. (1991). A comparison of input-output and social accounting methods for analysis in agricultural economics. In: Midmore, P. (ed.) *Input-output models in the agricultural sector*, Aldershot: Gower, 31-47.

Roberts, D. (1992). UK agriculture in the wider economy: an analysis using a social accounting matrix, unpublished PhD Thesis, Department of Agricultural Economics, University of Manchester.

Roberts, D. (1994). A modified Leontief model for analysing the impact of milk quotas on the wider economy, *Journal of Agricultural Economics*, 45, 90-101.

Round, J.I. (1989). Decomposition of input-output and economy-wide multipliers in a regional setting. In: Miller, R.E., Polenske, K.R. and Rose, A.Z. (eds.) *Frontiers of input-output analysis*, New York: Oxford University Press, 103-108.

Skolka, J. (1989). Input-output structural decomposition analysis for Austria, *Journal of Policy Modelling*, 11, 45-66.

Stone, R. and others (1962). *A computable model of economic growth. Volume 1 of a programme for growth*, Cambridge and London: Department of Applied Economics (University of Cambridge) and Chapman and Hall.

Szymer, J.M. (1992). Input-output coefficients and multipliers from a total flow perspective, *Environment and Planning A*, 24, 921-937.

Tyler, G. (1966). The use of multiple regression analysis of whole farm data in the estimation of enterprise labour coefficients, *The Farm Economist*, 11, 106-119.

Chapter two: Agriculture's Links with the Rural Economy: An Input-Output Approach?

Lucy Harrison-Mayfield
Department of Agricultural Economics and Management, The University of Reading

Introduction

Close inspection of recent writing on the rural economy reveals a set of attitudes concerning agriculture's links with the rest of the rural economy. Though these attitudes are sometimes contradictory, they do possess one common characteristic – each appears to be based more on 'gut' feeling than on any substantial evidence. Some examples follow:

> Links between town and hinterland are now less concerned with the marketing of produce and the interlinkages of goods and animals, and more with the use of facilities of the town by people living in the country.
>
> (Green, 1971: 6)

> Even in the agricultural industry, much of the supply trade (manufacture or repair of machinery or production of fertiliser and pesticides), is no longer dispersed throughout rural communities, but concentrated in a few urban sites.
>
> (European Commission, 1988)

> (*In the event of a drop in agricultural support*) . . . there will be a notable knock-on effect on employment in the support industries, many of which are rural.
>
> (National Economic Development Council, 1989: 6)

> Even in the most remote rural areas, the proportion of the population engaged in or connected with agriculture these days is rarely more than 15 to 20 per cent. This means that even in the most agricultural parts of Britain, 80 per cent of the

rural population will have no connection with agriculture whatsoever.

(House of Lords, 1990: 21)

The difficulty with multiplier effects is that increasingly, because of the way in which the food industry has changed – not just the farming but the food industry, those multiplier effects are not pinned down in rural areas ... therefore local multiplier effects from agriculture have diminished.

(House of Lords, 1990: 24)

These statements highlight a gap in knowledge which undoubtedly hinders full understanding of the way the rural economy operates. This is especially important at the moment, with so much discussion over the future of rural areas and, in particular, over agriculture's role within them. Nowhere is there a complete picture of the rural economy and its links with agricultural production. Do farmers trade locally, or does all trade now take place in the urban centres? Which industries have particularly strong connections with agriculture, and where are they located? Will a reduction in agricultural production necessarily affect other rural employment? These are essentially the types of question circulating in the policy arena that a rural input-output model could explore. Agriculture's role in providing rural employment indirectly through its links with other *rural* industries has not yet been established, and such a model could contribute to our understanding of the spatial distribution of farmers' inputs and outputs, and thus throw light on how much rural employment is indirectly related to agriculture.

Input-output analysis has been used to determine the employment indirectly associated with agriculture (see, for examples, Errington, 1986; Midmore, 1987; Errington and Harrison, 1990; and Leat and Chalmers, 1991), and would appear on the face of it to be an ideal tool to examine how much of such employment is rural. But is this a realistic assumption? This chapter aims to explore the difficulties with such a model, and to discuss ways of overcoming them.

A rural input-output table

A rural input-output table would give useful information by exploring the flows of goods between urban and rural areas, taking the form of an inter-regional table as Miernyk (1967) described. These differ from tables that describe single regional economies that aggregate external dependencies as primary inputs, in that they include the inter-regional as well as the inter-industrial independencies: or, in this case, inter-urban/rural independencies.

Most governments now compile national input-output tables of some sort. The UK tables are based on periodic surveys and, in recent years, have been published at approximately five-yearly intervals (the most recent being for 1990:

Central Statistical Office, 1995). Tables can then be constructed for regions either by mechanically reducing input-output coefficients from national tables, by survey methods (specifically focusing on rural input-output relations and the rural and urban sources of inputs), or more usually a combination of these two approaches.

Ideally, a compromise could be attempted for a rural table. The data collected for previous research that identified the location of farmers inputs and outputs (Harrison, 1993) could be used to estimate the agricultural coefficients for the rural-rural matrix: the coefficients for the urban-urban agricultural sector might then be derived by subtraction. Estimates of total manufacturing output for each district in the UK could be obtained from the Business Statistics Office, and categorised as rural or urban respectively. To derive the input-output coefficients for the other sectors, coefficient reduction methods are available, such as Simple or Cross Industry Location Quotients. Both techniques have low computational and data requirements. The methods have been described in Richardson (1972) and Miernyk (1967); their use in a rural economic context has been demonstrated by Johns and Leat (1987) in their study of the Grampian region.

To provide consistency in the resulting matrix, a matrix balancing technique can be used. Probably the most straightforward technique is the RAS method, described for example in O'Connor and Henry (1975). However, as the next few pages discuss, there are considerable difficulties with this approach. This chapter aims to bring to light the difficulties involved in applying these approaches, and to discuss means of overcoming them. The main problems can be divided into three different types; theoretical problems, national data availability, and survey data collection.

Theoretical problems

The first theoretical difficulty is 'what is rural?' This question has been discussed at length elsewhere (see Errington, 1989; Harrison, 1992) without conclusion, and it is inevitable that use of any classification will be controversial. Nevertheless, the choice of division between urban and rural will depend on the use to which it will be put, and compromises need to be made between what is ideally required and the ease at which it can be put into practice. That leads, in many cases, to the adoption of existing classifications, such as those of Cloke (1977), or Craig (1987).

Another difficult theoretical problem for creating a rural input-output table – which also applies to the creation of a regional table, although is more exaggerated in this instance – is that of estimating the flows of imports and exports. Any part of a nation is likely to be less self-sufficient than the nation itself and therefore require a greater percentage of imports. This is even more the case with regard to rural areas as the rural region is not a continuous region, so trade in and out will be much greater and more diverse than in other potential regions. Yet since no trade figures between rural and urban areas exist, unless a costly and time-consuming survey is undertaken, other means of inferring these problems will remain.

The Cross Industry Location Quotient (CILQ) coefficient adjusting technique makes some attempt at estimating where possible shortfalls in self-sufficiency may occur, and where imports into the region are therefore necessary. This is done in the following manner. The CILQ is equal to the proportion of producing industry *i* in the region compared to the national figure (or rural area), divided by the proportion of purchasing industry *j* in that region.

$$CILQ_{ij} = \frac{E_i^r / E_i^n}{E_j^r / E_j^n} \qquad [2.1]$$

where E_i^r is the level of employment in industry *i*, in region *r* (as an approximation for the level of output); the superscript *n* indicates employment in the whole national economy. From this, we can infer whether the selling industry will be able to provide enough goods to the purchasing industry. This method overcomes the initial problem of assuming that the demand patterns are the same, as it takes into account that the regional coefficients are not necessarily equal to the national coefficients, and that the trading potentials of the each industry vary between regions. It does not, however, consider the weighting of the two industries relative to the total output of the region.

If the Cross Industry Location Quotient is equal to one, there is the same proportion of national employment of each industry *i* and *j* and the national coefficient is left unadjusted. If the Cross Industry Location Quotient is greater than one, then the coefficient is again unchanged, since the proportion of industry *i* in the region is greater than industry *j*, and the assumption is that *i* can provide all the inputs required by industry *j*. On the other hand, if the Cross Industry Location Quotient is less than one, industry *i* is considered not to be producing enough in that region to supply industry *j* and imports would be required to make up the short-fall in demand. In this case, the coefficient is reduced by multiplying with the Cross Industry Location Quotient.

The fact that this method does not consider the weighting of the two industries relative to the total output of the region may create a problem for a rural input-output table – where the industrial structure of such a region is markedly different to that of an urban area, in a way that a regional analysis including towns would not. Agriculture is far more important to employment in rural areas, and this would not be recognised using such a technique. Hence in this instance the Simple Location Quotient (SLQ) for adjusting coefficients may be more appropriate. This is given by the following:

$$SLQ_i = \frac{E_i^r / \sum_{i=1}^{s} E_i^r}{E_i^n / \sum_{i=1}^{s} E_i^n} \qquad [2.2]$$

The Simple Location Quotient thus measures the relative importance (in terms of percentage of total employment) of the industry in the region compared to the nation.

National data availability

There are four main problems related to the use and potential availability of national data:

1. the availability of data at a small enough regional level for it to be classified as 'rural' or otherwise;
2. the possibility of sufficient disaggregation of data, in terms of the Standard Industrial Classification (SIC) to reveal details of the sectors on which particular interest is focused (for example, disaggregation of SIC division 4, other manufacturing, is required to highlight the food processing sector);
3. the availability of the data in the year that is to be the study year;
4. the non-availability of certain types of data.

Each is discussed in turn in the following sub-sections.

Regional level of data

Industrial output data are vital to determine the gross output vector at a rural level. The Business Survey Data (Central Statistical Office, 1992) are only available widely on a broad regional level, and at a cost, at a district level. The districts could be categorised into urban and rural areas (perhaps using Cloke's typology), thus producing an estimate of the 'rural' output. This would restrict the definition of 'rural' to one based at district level. However, employment data are available at a ward level and this can be used to estimate output. This is of course not the best option - employment data are never a good substitute for output data, but may help overcome other problems which emerge in the following section.

Aggregation of industrial classification

The means of aggregation of industrial classification in data sets turns out to be highly restrictive. Input-output tables produced by the Central Statistical Office divide industries into 102 classes, separating those relating to food and drink, but compressing all services into a single category. However, the rural employment statistics (from the Population Census) are available in 60 classes, which are differently aggregated and combine all the Food, Drink and Tobacco industries into a single category, though service sectors are more disaggregated. For instance, taking the 1984 input-output tables (Use Matrix) from Central Statistical Office (1988) as a starting point for such a model, the original 102 industrial sectors need to be aggregated to 42 in order to balance the rural employment data industrial categories with those used in this table. This is necessary owing to the data available but causes many important agriculturally-related sectors to disappear (for example, agricultural machinery is subsumed in Mechanical Engineering, and all the separate food industries come under the heading Food, Drink and Tobacco), detracting from the potential usefulness of the resulting input-output table.

Business Statistics Office output statistics (Central Statistical Office, 1992) aggregate at an even higher level, into 11 groups, and amalgamate all production industries together, although they also separate retail and wholesale distribution, and various services. Although it is possible to determine which activities are in each division, the result is a huge loss of information, as the number of industrial sectors which can be dealt with is reduced to five. Hence despite the inaccuracies that would no doubt be incurred, it may be worth using employment data as a proxy for output, as not only are they available at ward level, but they enable some of the more relevant industrial sectors to be distinguished: the greater level of detail might also offset aggregation bias, discussed seminally by Theil (1957).

Frequency of availability of data

The largest restriction in this case is the publication of the national input-output tables themselves. They are not only published fairly infrequently, but by the time they are issued have invariably become rather out-of-date. The most recently published UK tables are for 1990 (Central Statistical Office, 1995). Since the adoption of the input-output table as a means of checking national accounts control totals, the frequency of estimation may increase (though this may have no effect on the frequency of publication, of course, it may mean that the tables can be made available to researchers on an informal basis). Employment data from the Population Census are only available every ten years, though the Census of Employment is undertaken more frequently, and the changes in trends drawn from these data can be used to adjust the more comprehensive information from the Population Census.

The non-availability of data

The data which would be very useful to have are separate industrial employment production functions. In order to examine the employment implications of this study, it is necessary to evaluate the relationship between the output of firms and their employment. Ideally, this relationship could be expressed by an employment production function, using simple linear regression techniques of the form:

$$E_i = a + bX_i + u_i \qquad [2.3]$$

where E_i is the employment of each industry, and X_i is the gross output of the industry. Unfortunately, data to estimate this function are not readily available, so a point approximation of the form:

$$E_i = bX_i \qquad [2.4]$$

based on a single pair of observations is normally used. This may tend to lead to an overestimation of the employment effects of any change in output. Employment production functions may also differ between urban and rural areas, and this could be a useful study to pursue, bearing in mind the many estimations of employment change that are based on very crude calculations.

Data collection

Examining agriculture's links with the rural economy implies a particular interest in the spatial distribution of the agricultural industries' purchases and sales. This sector is therefore a prime candidate for the capture of superior data, rather than mechanically adjusted coefficients. This section illustrates problems of obtaining such data by describing a recent attempt to adjust the agricultural coefficients. The data were obtained from the Reading province of the Farm Business Survey. A sample of farm accounts was analysed, and from information found in the invoices and receipts, it was possible to obtain postcodes of firms involved in the transactions; this proved to be an efficient method for deducing the location of firms dealing with the farm.

The first stage of the analysis process sought to discover whether the postcodes were located in rural or urban wards. The rural classification used was that of Craig (1987) and based essentially on the population density and land use of each ward. Each ward in England and Wales has been classified by Craig's method on a scale running from 1 (wholly urban) to 6 (wholly rural). Combining database information allocating postcodes to wards, and wards to Craig's categories, it was possible to attach the relevant level of rurality to each postcode, and each transaction.

There are two problems relating to this data, however: firstly in coding (classification of product, destination/source of product); secondly, in the sample size.

Coding

The information required from each invoice and receipt was the type of product, the value of the transaction and the postcode of the firm that bought or sold the good.

a) The classification of the product

This data collection began initially as an investigation into farmers' first round links with other industries, with no real thought of construction of a rural input-output table. The first difficulty was whether the product should be classified by type of product, or by the industry from whence it came. This is a classic problem where input-output tables are concerned, since many industries produce secondary products or a range of products. As a starting point, the classification used by Farm Accounts conventions was used, based on the type of product, but this created complications when aiming to fit the figures into the Standard Industrial Classification required by an input-output matrix.

b) Destination/source of product

There were several decisions to be made about issues regarding the destination or source of product (e.g. relating to Co-ops, missing postcodes, invoices from central offices, etc.). The most difficult of several was determining what the policy should be when the address was a distributor rather than a producer of the good in question: this was the case with many of the inputs, most especially fertilisers and sprays. For example, it was very obvious that most of the sources of fertiliser were not actually manufacturing it themselves, but merely selling it along with other farm products. This distinction between the retail and the distribution of goods is not made in the classification because it was not always possible to distinguish between the two. This is a major drawback to the study, since it inevitably concentrates on the first round of transactions between farms and the rest of the economy. One possible means of overcoming this in future studies would be to undertake a survey of the distributors to find out where they purchased the inputs, or to survey the original producers to discover the volume sold through each of the various channels. This is effectively the way this problem is dealt with in the compilation of the national input-output tables, where gross margins from the inquiries into the Wholesale and Retail trades are used to reduce purchasers' prices on the demand side to producers' values (Central Statistical Office, 1988).

Size of sample

The Farm Business Survey Investigational Officers collect accounts from farms and are usually given about one week to work on them, before returning them.

Five Investigational Officers agreed to pass on ten accounts each, but this did mean that; 1) there was a very limited sample size; 2) the time available to analyse them was very limited, usually between 24 and 48 hours; and 3) this process involved no random 'sampling' and as a result, there are likely to be systematic biases in the sample. When the final sample was examined, it was discovered that there were more small farms than in the Reading Farm Business Survey sample, and also more dairy, but fewer cereals and cattle and sheep farms. It should be noted that, since all the farms were inside the Reading Farm Business Survey province, which stretches from the West Country, down to the South Coast and as far east as Slough, i.e. the majority of central southern England, the data extracted in this exercise cannot be assumed to be representative of the farming industry of the UK as a whole.

Results

The subsequent analysis of these data revealed some interesting results, particularly those relating to the linkages of large and small farms. Some illustrative results are shown in Tables 2.1 to 2.4, and show the distribution of transactions by number, and by value in each category for farm inputs and outputs. The number of transactions were thought to be important in that small frequent purchases might indicate a stronger link than a large one-off purchase. However, the transaction figures are based in some instances on bills relating to a month's or a quarter's purchases or sales, and so may underestimate the true number of dealings with a firm.

Tables 2.1 and 2.2 show the distribution of inputs and outputs between urban and rural areas. The pattern that emerges is of small farms having more frequent transactions and buying a greater value of goods in the more rural areas (Mixed Urban-Rural, Predominantly Rural and Rural) while the larger farms tend to have a greater percentage of their transactions with the Urban and Predominantly Urban areas and to purchase a higher value of goods from these areas.

Table 2.1. Comparing the 'ruralness' of sources of inputs between small and large farms

	Percent. Transactions		*Percent. Value*	
Source of Input	*Large*	*Small*	*Large*	*Small*
Urban	35.2	24.3 ***	27.4	21.6
Predominantly Urban	21.7	17.6	19.0	13.7
Mixed Urban-Rural	26.1	30.6	32.9	42.9 **
Predominantly Rural	5.4	17.8 *	9.4	14.1 ***
Rural	5.7	4.3	5.4	6.0

Statistical significance: * ~ 1%; ** ~ 5%; *** ~ 10%.

Table 2.2. Comparing the 'ruralness' of destinations of outputs between small and large farms.

Destinations of Outputs	Percent. Transactions Large	Percent. Transactions Small	Percent. Value Large	Percent. Value Small
Urban	41.0	30.6 **	34.0	17.5 **
Predominantly Urban	20.4	9.5 *	19.5	12.8
Mixed Urban-Rural	17.1	36.4 **	32.3	54.5 *
Predominantly Rural	10.6	14.5	5.9	8.4
Rural	3.3	3.7	1.7	6.7 ***

Statistical significance: * ~ 1%; ** ~ 5%; *** ~ 10%.

The results comparing the distances over which inputs and outputs are bought and sold respectively are shown in Tables 2.3 and 2.4. These show a less substantial difference between small and large farms, but do indicate that more of the farm inputs and outputs for the small farms are being bought and sold within a 10 kilometre radius than for the large farms, both in terms of the total number and value of transactions. There is, however, a very marked difference in the value of farm output, as Table 2.4 shows. For the small farms, 63 per cent of the output is shown to be distributed within 5 kilometres of the farms, compared to only 21 per cent of the output for the large farms. These results are mostly expected as the large farms will be able to take advantage of economies of scale, enabling them to transport their produce over a greater distance in search of better prices, or purchase inputs in bulk, perhaps straight from the factory.

Table 2.3. Comparing the distances over which inputs were bought.

Distance from Farm (km)	Percent. Transactions Large	Percent. Transactions Small	Percent. Value Large	Percent. Value Small
0 - 4.9	47.6	53.0 ***	37.5	41.5
5 - 9.9	14.7	14.9	15.7	17.0
10 - 19.9	12.3	12.0	10.7	11.4
20 - 49.9	12.9	11.1	15.6	9.2 **
50 or more	12.5	8.9 ***	20.6	21.3

Statistical significance: * ~ 1%; ** ~ 5%; *** ~ 10%.

Table 2.4. Comparing the distances over which outputs were sold.

Distance from Farm (km)	Percent. Transactions Large	Percent. Transactions Small	Percent. Value Large	Percent. Value Small
0 - 4.9	33.0	41.0 **	21.0	63.0 *
5 - 9.9	15.0	13.2	22.0	5.0
10 - 19.9	21.0	12.0	24.0	7.0 ***
20 - 49.9	21.0	14.9	11.0	8.0
50 or more	10.0	20.7	24.0	17.0 *

Statistical significance: * ~ 1%; ** ~ 5%; *** ~ 10%.

These findings are important in their own right – the demonstration of rural links illustrates the importance of small farms to the rural economy. It supports the case for targeting public support to small farms, since the disappearance of small farms would have greater impact, not only through direct employment effects but also through the indirect employment associated with the local linkages. This analysis has particularly important implications for regions where farming is predominantly based on small enterprises, especially as these tend to be the more marginal and remote areas, often with other, related, social and economic difficulties. The demonstration of close local links between small farms and their immediate vicinity shows how the demise of small farm would have an impact at a very localised level, thus leading to a spatial concentration of recessionary effects.

These results indicate a possible source of superior data to improve estimates of spatially disaggregated linkages between agriculture and the local/rural economy, though it would be preferable if they were confirmed by replications throughout the UK.

An illustrative example

This chapter serves to illustrate the practical difficulties that occur in creating input-output tables, and demonstrates the nature of time and data resources required to compile them in a realistic fashion. However, although there are other ways of estimating 'first round' links between agriculture and the rural economy (see Harrison, 1993) that tend to emphasise the direct importance of agriculture to the local rural economy, there are no doubt considerable indirect impacts that are felt through the rural economy resulting from changes in farm production. These can only be effectively represented by using input-output modelling. To demonstrate this, the 1984 UK input-output table (Use Matrix) from Central Statistical Office (1988) has been adjusted mechanically to produce a very crude rural input-output table. The sectors were aggregated to 42 to comply with the rural employment data (Population Census 1981, adjusted to 1984). The coefficients were adjusted using Cross Industry Location Quotients, and the output and employment multipliers derived from the resulting input-output table are shown in Table 2.5. Agriculture's employment multiplier is the fourth largest in the table, after Coal and Coke extraction, Other Vehicles and Food, Drink and Tobacco. The other multipliers greater than two are Electricity, Leather Goods, Construction and Insurance. Insurance seems especially surprising as service industries usually have low multipliers. While little reliance can be placed on results from a mechanically produced table, it is interesting to note that the agricultural multipliers are very similar to those produced by Hubbard (1982) in his analysis of the West Durham region. Perhaps this is not surprising, since he also used Cross Industry Location Quotients to adjust an earlier national table.

Conclusion

Data availability severely limits the possibilities of producing a realistic rural input-output table, since the time and cost of collecting all the supplementary data required would be prohibitive. The spatial disaggregation of the existing data is not adequate for use in conjunction with user-defined 'rural' regions (as opposed to existing government administrative areas), as in order to build up one's own 'region', one needs building blocks of wards for which the data are very limited. To compound these difficulties, many of the sources do not publish data at a sufficiently disaggregated industrial level to enable a distinction between the industries which are of particular interest, or with any attempt to make the aggregations compatible with other data sources. To make the data adequate would require firstly larger sample sizes (to enable wider publication of ward level data) and secondly greater industrial and spatial disaggregation of published statistics.

Given these difficulties, it may be more sensible to pursue this research through 'case studies' of particular localities rather than attempting to model the whole rural economy. The case study described in this chapter emphasises the importance of small farms to the rural locality. Other studies could use the methodology outlined here to explore, for example, the importance of specific towns to the surrounding rural areas, or the role of pluriactivity in the rural economy. There is thus certainly scope to produce small-scale input-output models which would add to the understanding of agriculture's contribution to this rural economy. These studies could produce useful illustrative examples of the workings of the rural economy until such a time when data are more widely available in a suitably disaggregated form for a national level model to be developed.

Table 2.5. Output and employment multipliers for rural UK.

Sector	*Output Multiplier*	*Direct Employment Coefficient*	*Direct & Indirect Employment Coefficient*	*Employment Multiplier*
Agriculture	1.26	23	51	2.22
Forestry & Fishing	1.12	23	46	1.99
Cole, Coke, etc.*	1.63	124	290	2.34
Extraction of Oil, etc.	1.12	1	1	1.24
Oil Processing	1.86	1	1	1.37
Electricity	1.67	14	29	2.07
Gas	1.60	14	15	1.05
Water	1.23	37	54	1.45
Extraction of Metals	1.16	2	4	1.79
Metal Manufacturing	1.81	18	36	1.98
Extraction of Minerals	1.23	39	76	1.95
Non-metals	1.33	32	57	1.77
Chemicals	1.52	17	31	1.84
Man-made Fibres	1.56	19	34	1.79
Metal Goods	1.68	38	73	1.91
Mechanical Engineering	1.51	27	51	1.88
Office Engineering	1.57	34	55	1.63
Electrical Engineering	1.57	32	53	1.67
Motor Vehicles	1.05	27	29	1.06
Other Vehicles	1.95	36	95	2.65
Instrument Engineering	1.21	49	62	1.26
Food, Drink & Tobacco	1.72	18	41	2.28
Textiles	1.59	41	69	1.69
Leather	1.57	34	71	2.09
Clothing	1.23	58	75	1.29
Timber	1.42	39	71	1.83
Paper	1.53	31	53	1.71
Rubber	1.57	31	54	1.74
Other Manufacturing	1.46	37	63	1.70
Construction	1.64	23	48	2.07
Distribution	1.46	52	90	1.73
Hotels	1.29	88	142	1.61
Rail	1.96	72	137	1.90
Road & Inland Transport	1.36	48	80	1.66
Sea Transport	1.52	10	16	1.59
Air Transport	1.55	11	19	1.70
Other Transport	1.47	51	80	1.56
Post/Telecom	1.29	42	60	1.43
Banking	1.57	23	37	1.63
Insurance	1.87	26	53	2.04
Other Services	1.23	76	109	1.45

* May be distorted due to Miners' Strike in 1984.

References

Central Statistical Office (1988). *Input-output tables for the United Kingdom, 1984*, London: HMSO.

Central Statistical Office (1992). *Business monitor PA1003 (1992) size analysis of UK businesses*, London: HMSO.

Central Statistical Office (1995). *Input-output tables for the United Kingdom 1990*, London: HMSO.

Cloke, P.J. (1977). An index of rurality for England and Wales, *Regional Studies*, 11, 31-46.

Craig, J. (1987). An urban-rural categorisation for wards and local authorities, *Population Trends*, 32, 6-11.

Errington, A.J. (1986). Employment. In: Craig, G.M., Jollans, J.L. and Korbey, A. (eds.), *The case for agriculture: an independent assessment*, CAS Report No 10, Reading: Centre for Agricultural Strategy (University of Reading).

Errington, A.J. (1989). Measuring the employment structure of rural Britain: the basis for a rural development strategy, paper presented to the Annual Agricultural Economics Society Conference, Aberystwyth, 7-10 April 1989.

Errington, A.J. (1991). Input-output relationships and the rural economy. In: Midmore, P. (ed.) *Input-output models in the agricultural sector*, Aldershot: Gower, 40-56.

Errington, A.J. and Harrison, L. (1990). Employment in the food system. In: Marsden, T. and Little, J. (eds.) *Political, social and economic perspectives on the international food system*, Aldershot: Avebury, 115-140.

European Commission, (1988). *The future of rural society*, Bulletin of the European Communities Supplement 4/88, Brussels: European Commission.

Green, R. (1971). *Country planning: the future of rural regions*, Manchester: Manchester University Press.

Harrison, L. (1992). Agriculture's role in rural employment; a study of the relationship between agriculture and other rural employment in England and Wales, unpublished PhD Thesis, Department of Agriculture, University of Reading.

Harrison, L. (1993). The impact of the agricultural industry on the rural economy – tracking the spatial distribution of the farm inputs and outputs, *Journal of Rural Studies*, 9, 81-88.

House of Lords (1990). *The future of rural society*, 24th report Select Committee on the European Communities, Vol. 1: Report. London: HMSO.

Hubbard, L. (1982). *New jobs in West Durham: an ex ante study of the Development Commission's advance factory programme*, Agricultural Adjustment Unit Monograph No. 9, Newcastle upon Tyne: Department of Agricultural Economics (University of Newcastle upon Tyne).

Johns, P.M. and Leat, P.M.K. (1987). The application of modified GRIT procedures to rural development analysis in Grampian region, *Journal of Agricultural Economics*, 38, 243-256.

Leat, P.M.K. and Chalmers, N. (1991). Analysing agricultural adjustment in Grampian using an input-output model of the agricultural and food complex. In: Midmore, P. (ed.) *Input-output models in the agricultural sector*, Aldershot: Gower, 57-71.

Midmore, P. (1987). *The impact of farm prosperity on the rest of the Welsh economy*, discussion paper, Aberystwyth: Department of Agricultural Economics and Marketing (University of Wales).

Miernyk, W.H. (1967). *The elements of input-output analysis*, New York: Random House.

National Economic Development Council (1989). *Work in the countryside: agriculture and rural employment in the 1990s*, London: The National Economic Development Council's Agriculture Ad-hoc Sector Group.

O'Connor, R. and Henry, E.W. (1975). *Input-output analysis and its applications*, London: Charles Griffin & Company.

Richardson, H.W. (1972). *Input-output and regional economics*, Trowbridge: Wiedenfeld and Nicolson, Redwood Press Ltd.

Theil, H. (1957). Linear aggregation in input-output analysis, *Econometrica*, 25, 111-122.

Chapter three:
Measuring the Regional Economic Impact of Pluriactivity on Scottish farms

Morag Mitchell
Economics, Marketing and Management Department, The Scottish Agricultural College, Auchincruive

This chapter outlines some of the work carried out by the Pluriactivity in the Agricultural Sector in Scotland project, funded by the Economic and Social Research Council as part of the Joint Agriculture and the Environment Programme (JAEP), and describes how the regional economic impacts of a farm household setting up on-farm non-agricultural enterprises were modelled. This included, firstly, reducing national input-output tables to a regional level; secondly, estimating the income and expenditure flows of various enterprises; and, finally, using a gravity model within a Geographical Information System (GIS) to spatially distribute and present the employment and income impacts of various enterprises within three regions of Scotland – Dumfries and Galloway, Grampian and Fife. The results (described in moe detail in Mitchell and Doyle, 1993) show that the size and distribution of the economic impacts of different enterprises vary, depending on where they are located. These are supported by maps that show the spatial distribution of these impacts when a caravan site is established in various locations across Grampian region.

Regional input-output tables

The first step was to produce regional input-output tables, which were used to estimate the upstream and downstream effects of establishing an on-farm non-agricultural enterprise on all sectors of the economy, not just agriculture. In Scotland, national, but not regional, input-output tables were produced for 1989 (Scottish Office Industry Department, 1994). Therefore a method was adopted to reduce these national tables to regional ones. The cost of using survey methods to create regional input-output tables was prohibitive and, as a result, the Generation of Regional Input-Output Tables (GRIT) method developed by Jensen *et al.* (1979) and described by Johns and Leat (1986) was used to produce a regional coefficient matrix, which in turn was used to calculate income and employment multipliers. The rest of this section outlines this methodology: a more detailed discussion may be found in Jensen *et al.* (1979).

The national table includes elements of the Standard Industrial Classification (SIC), which were reduced to ten single digit SICs by merging cells. This drastic reduction in detail was necessitated since the employment data used in coefficient estimation were not available for the regions in any less detail. Also, because the final results were to be presented at a parish level, in many respects the presentation of results in such detail would have been ineffectual. Table 3.1 outlines the ten single digit SICs used throughout.

Table 3.1. Single digit Standard Industrial Classification divisions.

	Description of SIC
Division 0	Agriculture, forestry, fishing
Division 1	Energy and water supply
Division 2	Extraction of minerals/ores, manufacture of metals, mineral products and chemicals
Division 3	Metal goods, engineering, vehicles
Division 4	Other manufacturing
Division 5	Construction
Division 6	Distribution, hotels, catering
Division 7	Transport, communication
Division 8	Banking, finance, insurance, business services
Division 9	Other services

A regional input-output coefficient matrix for each of the three study areas was then estimated from the aggregated national coefficient table, using a matrix of Cross Industry Location Quotients (CILQ), which compare the employment rates in each industry in the region to the Scottish level. CILQs are given by the equation:

$$CILQj = \frac{E_i^r / E_i^N}{E_j^r / E_j^N} \qquad [3.1]$$

where E_i^r refers to numbers employed in the *i*th industry in region *r*, E_j^r refers to numbers employed in the *j*th industry; and, respectively,the superscript *N* refers to total numbers employed nationally

If any value in the CILQ matrix lay between 0 and 1 it was assumed that the production in the region within that industry was insufficient to meet regional demand, and imports were required. In that case, the regional input coefficients were arrived at by multiplying the national coefficient by the CILQ. If, however, the CILQ was greater than 1, regional production was assumed to be sufficient to meet regional demand and the input-output coefficients of the Scottish matrix were not reduced in the local matrix.

The regional input-output coefficient matrices formed in this way were used to calculate the Type I and Type II income and (drawing also on the employment data originally used to reduce the coefficients) employment multipliers. An increase in output in one industry will require inputs from other sectors of the economy, hence there is a knock-on effect and the Type I multipliers describe these direct and indirect impacts. The direct and indirect increases in output will, by increasing incomes, induce further multiplication of income and/or employment impacts and the total direct, indirect and induced effects are given by the Type II multipliers. Table 3.2 shows the Type II income and employment multipliers calculated for each of the three regions.

Table 3.2. Regional Type II income and employment multipliers.

		Fife		*Grampian*		*Dumfries & Galloway*	
	Division	*Inc.*	*Emp.*	*Inc.*	*Emp.*	*Inc.*	*Emp.*
Agriculture, forestry & fishing	0	3.511	2.941	2.529	1.834	1.803	1.437
Energy & water	1	1.671	1.890	1.383	1.518	2.527	3.071
Extraction of minerals/ ores, manufacture of metals, mineral products & chemicals	2	1.834	2.542	3.395	5.306	2.047	2.910
Metal goods, engineering & vehicles	3	1.253	1.332	1.579	1.765	1.301	1.382
Other manufacturing	4	1.414	1.482	1.676	1.815	1.423	1.529
Construction	5	1.889	1.451	2.008	1.540	1.436	1.226
Distribution, hotels & catering	6	1.389	1.314	1.475	1.393	1.818	1.657
Transport & communication	7	1.535	1.693	1.534	1.705	1.376	1.487
Banking, finance, insurance & business services	8	1.365	1.331	1.450	1.416	1.194	1.173
Other services	9	1.208	1.234	1.329	1.374	1.129	1.139

It can be seen from the table that the Type II income multipliers for Grampian region tend to be higher than Dumfries and Galloway or Fife. This may imply that Grampian has much stronger inter-industry links, relying less heavily on imports than Fife or Dumfries and Galloway. Therefore, an increase in output will impact more on income within Grampian than within the other two.

Fife also has relatively low Type II employment multipliers in some sectors, indicating that an increase in output does not have a significantly large knock-on effect within the region. This may be due to Fife being small in area, therefore employing smaller numbers of people within a sector. Also, the close proximity of Edinburgh and Dundee may mean that some of the impacts leak out of the region. Both Fife and Grampian regions exhibit their highest employment multipliers in divisions 0, 2 and 5, reflecting the importance of agriculture, the extraction of minerals/ores and construction to these economies, especially Grampian which has a heavy involvement in the oil industry. Dumfries and Galloway, on the other hand, has stronger employment multipliers in energy and water, extraction of minerals/ores and distribution, hotels and catering, reflecting a higher profile in the tourist industry. These comparisons enable distinctions to be made between how different enterprises will impact on the region.

There are limitations to using this method, besides those already mentioned, which need to be recognised:

1. There is an assumption that production functions are linear, and therefore a rise in output in one sector will create proportionately fixed demand for inputs from all other sectors, implying that input substitution or economies of scale are ignored. Whilst this may be acceptable for large national industries, it is less easy to justify when disaggregating to a regional level, or when the industry accounts for a very small proportion of the economy.
2. There is a closely related assumption that any increase in demand for inputs will be met by the supplying industries concerned through increased production, not merely by running down stocks or increasing imports from other regions.
3. Since an industrial sector is composed of a variety of products and enterprises which are aggregated to produce the input requirements and output flows of that sector. Therefore, using input-output tables to estimate the economic impact the establishment of a particular enterprise has on the economy assumes that it has the same pattern of inputs and outputs as all other enterprises within that sector.
4. Finally, there is no clue as to the time scale over which adjustments take place.

Nevertheless, because detailed survey work into links between the regional industries is so costly, the regional input-output tables produced by this methodology offer an approximate means of identifying the backward linkages of an enterprise, and have the advantage of being conceptually simple. However, results gained from them must necessarily be tentative and, before operational policy decisions are taken, should be supported by further research.

The multiplier effect on the on-farm enterprises

Turning to the on-farm enterprises, the variable and fixed costs of setting up and running representative sized businesses in the first year were estimated (Dalton and Wilson, 1989; MAFF, 1981; McInerney and Turner, 1991), and used to illustrate the methodology applied to distributing the economic impacts between SICs and calculating the multipliers for the various activities. This analysis can be extended to include subsequent years.

The expenditures incurred in the enterprise will increase the output of the SICs: the regional multipliers were used to calculate the overall impact of this increased output. Therefore, to be able to calculate the multipliers of the various enterprises, it was essential to know the regional multipliers for different sectors, and the different expenditure patterns of the various enterprises. This, of course means that some enterprises will have a greater impact on some regional economies than others.

Table 3.3 outlines the income multipliers for selected activities within the three regions. These income multipliers were not consistently higher in any one region illustrating that it may be more advantageous, in terms of aggregate effect, to establish particular enterprises in one region rather than another. The highest multipliers in Grampian and Fife were for farm shops, livery stables and shooting, whereas for Dumfries and Galloway they were caravan sites and bed and breakfast.

Table 3.3. Enterprise multipliers.

	Fife	*Dumfries & Galloway*	*Grampian*
Bed & Breakfast	1.389	1.818	1.475
Farm Shop	2.230	1.615	2.075
Caravan Site	1.652	1.816	1.672
Shooting	1.716	1.502	2.021
Livery Stables	2.206	1.582	2.057

The spatial distribution of the economic impact of pluriactivity

Having calculated the overall size of the income and employment effects of establishing the on-farm enterprises, their local spatial distribution was estimated using the gravity model of the SPANS Geographical Information System (GIS). This technique examines the potential effect of enterprise establishment in a particular geographic location, and the impact that it has on each parish within the region can be estimated (Tydac, 1991). The resultant spatial patterns may be used to determine whether an enterprise has a merely local, or wider regional impact.

The Gravity Model assumes that the probability of expenditure gravitating towards a parish is an indirect function of the distance between that parish and where the enterprise is established, and a direct function of its ability to attract some level of expenditure. Thus, the probability of expenditure being attracted to a particular parish is given by:

$$P_i = \frac{\alpha_i / (d_{ij})^2}{\sum_{i=1}^{j} \alpha_i / (d_{ij})^2} \qquad [3.2]$$

where P_i refers to the probability of expenditure being attracted to parish i, α_i is an attraction index and d_{ij} the distance between parish i, and parish j where the enterprise is based. The distances used were straight line distances from the centre of the parish where the proposed development will be established to the centres of all other parishes. The co-ordinates of the parish centres were generated by the GIS. The attraction index, α_i, reflects the existing SIC employment structure of the parish, weighted according to the expenditure pattern of the enterprise in question. Hence, it is assumed that the extra expenditure generated is attracted to the parishes which already have significant employment in the industries supplying the inputs to the enterprise. However, it is generally assumed in a model of this type (Cadwallader, 1981) that the attraction index will be the more important of the two; therefore the distance factor is squared so that changes in distance beyond the parish boundaries will have a smaller effect on the probability of that parish attracting some of the expenditure.

Results

The complete model produced maps which spatially distributed the economic impacts of selected on-farm enterprises. The following example shows the impact of a caravan site placed in different parishes in Grampian region. These impacts are stratified into four sections showing the cumulative percentage shares of the impact, starting with the parishes which benefit least.

For example, the lowest band, 0-8%, shows the parishes which gain the least or no benefit from the enterprise. These are cumulatively added until they reach 8%. The parishes which gain slightly more benefit are cumulatively added to 16% and the next level to 24%. The highest band, 24-100%, are the parishes which gain the most benefit. If the number of parishes in this top bracket is small, as it is with the caravan site in Fraserburgh or Belhelvie, it means that 76% of the economic benefits of the enterprise are very localised.

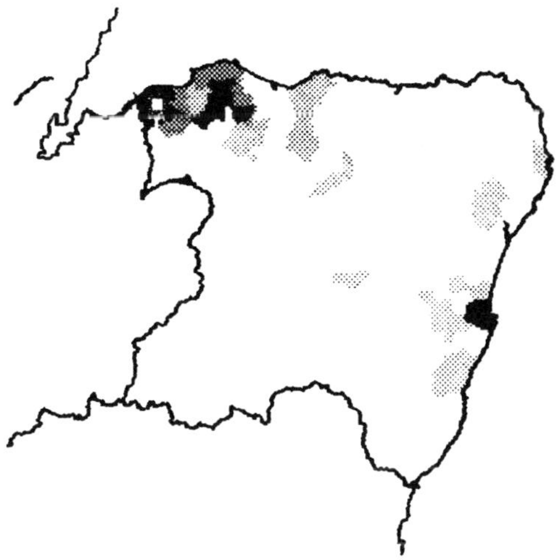

Dyke and Moy

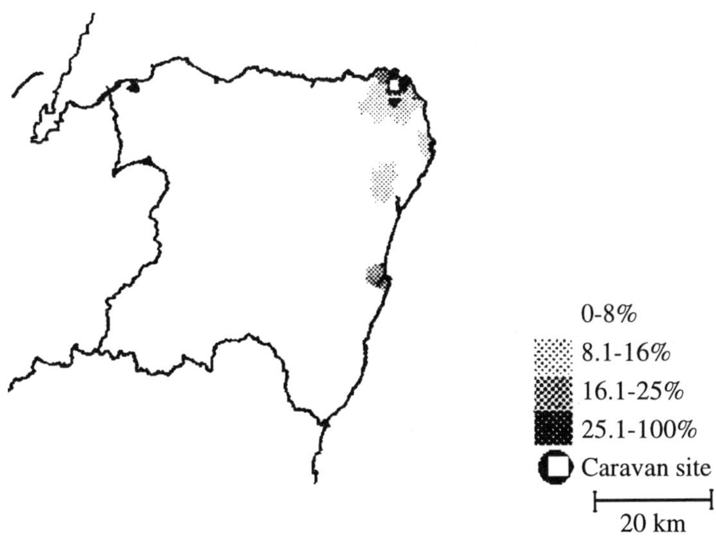

Fraserburgh

Fig. 3.1. (a) Economic impact of a caravan site placed in various parishes in Grampian region.

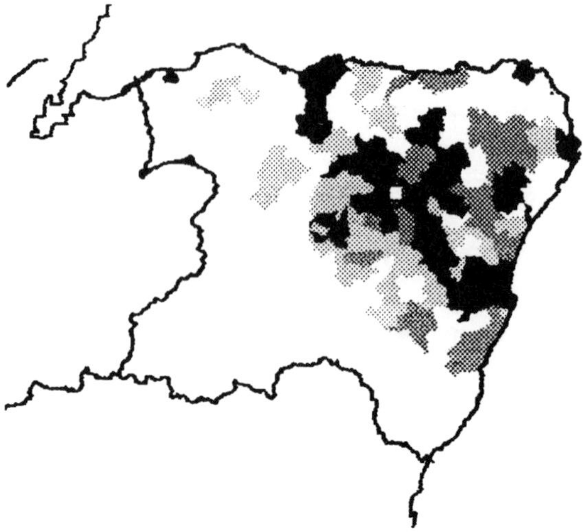

Insch

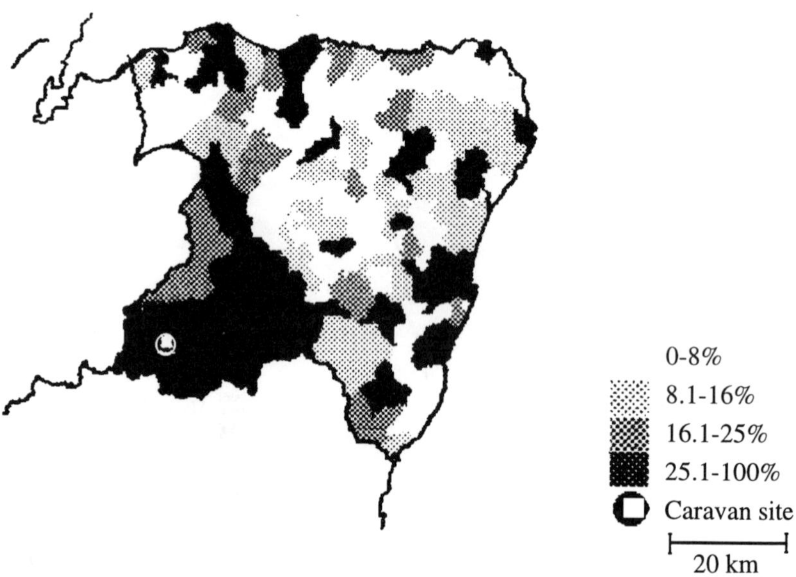

Crathie and Braemar

Fig. 3.1. (b) Economic impact of a caravan site placed in various parishes in Grampian region.

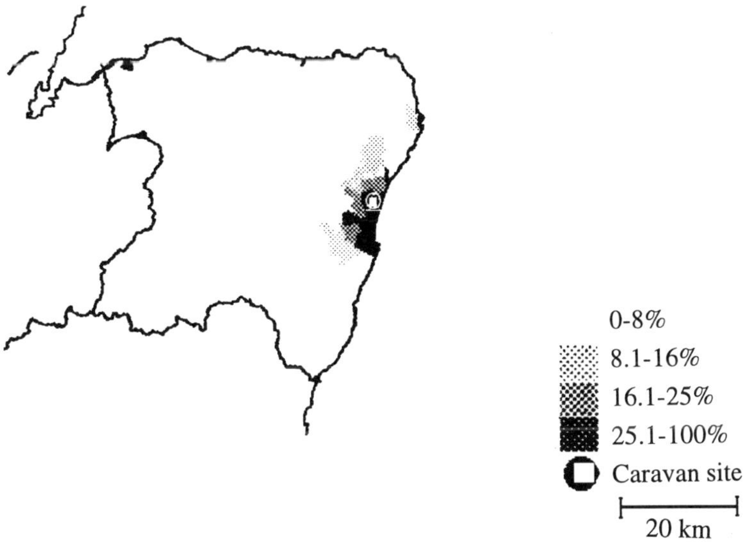

Belhelvie

Fig. 3.1. (c) Economic impact of a caravan site placed in various parishes in Grampian region.

The set of maps examines what happens if the same activity is moved round the region. Placing a caravan site in Fraserburgh, Dyke and Moy or Belhelvie means that there are large centres of population close by, especially in the case of Belhelvie which is on the outskirts of Aberdeen. Insch, on the other hand is comparatively rural and the impacts are spread around more parishes with no one centre being able to attract the major benefits. Crathie and Braemar show seven more spatially distributed effects right across the region. The south west of Grampian is a remote rural area including the Grampian Mountains and part of the Cairngorms so it is expected that it would not be economically active. The parishes do not have the ability to hold the economic benefits of setting up a caravan site, so it get spreads throughout the region.

Conclusions

In summary, although the techniques employed in this study were surrounded with considerable caveats, they do provide a way in which to estimate the spatial impacts of farm diversification on rural areas and open the way for discussion on whether regional policies should be used to target specific developments. These models can be used to illustrate the spatial distribution of the economic impact of various activities, and maps have been produced for a variety of enterprises in the three regions. The results suggest that the pattern changes depending on the type of enterprise, how close it is to a large town and the industrial structure of the region.

References

Cadwallader, M. (1981). A cognitive gravity model of consumer spatial behaviour, *Journal of Regional Studies*, 15, 279-284.

Dalton, G.E. and Wilson, C.J. (1989). *Farm diversification in Scotland*, SAC Economic Report No. 12.

Jensen, R.C., Mandeville, T.D. and Karunaratne, N.D. (1979). *Regional economic planning: generation of regional input-output analysis*, London: Croom Helm.

Johns, P.M. and Leat, P.M.K. (1986). *An approach to regional economic modelling: the case of Grampian*, NOSCA Economic Report No. 144.

MAFF (1981). *Direct farm sales to the public*, Socio-Economics Paper No. 8.

McInerney, J. and Turner, M. (1991). *Patterns, performance and prospects in farm diversification*, Exeter: Agricultural Economics Unit (University of Exeter).

Mitchell, M. and Doyle, C.J. (1993). *Pluriactivity on Scottish farms: regional socio-economic impacts and prediction of potential*, Report prepared for the Economic and Social Science Research Programme on Agriculture and the Environment, Edinburgh: The University of Edinburgh.

Scottish Office Industry Department (1994). *Scottish input-output tables for 1989*, Edinburgh: HMSO.

Tydac (1991). *SPANS 5.2*, Ontario: InterTydac Technologies Inc.

Chapter four: Methodological Issues in Forestry Input-Output Modelling

Kenneth J. Thomson[1] and Demetrios Psaltopoulos[2]

[1] *Department of Agriculture, The University of Aberdeen*

[2] *formerly Department of Agriculture, The University of Aberdeen: currently Department of Agricultural Economics, The Agricultural University of Athens*

Introduction

The application of the input-output technique (Leontief, 1951) to the analysis of the economic role of primary sectors, particularly in economic systems dependent on such activities, is not a new area of study. As noted in the first chapter, rural employment and income considerations, particularly on a regional dimension, have received considerable attention (Jensen *et al.*, 1979; Johns and Leat, 1986), relying on the technique's ability to identify and quantify inter-industry links (Miernyk, 1969). Similarly, Psaltopoulos and Thomson (1993) have reviewed forestry-centred applications at national and regional level, mostly concerned with the sector's output and employment-generating impact. This chapter explains and discusses some methodological issues arising from recent attempts by ourselves *et al.* to advance the use of input-output techniques in the analysis of afforestation policies in Scotland and Ireland (Gardiner *et al.*, 1994).

The anticipated transition of Western European agriculture towards liberalised market conditions is expected to result in gradual but serious decline of agricultural output and employment. This would significantly influence the economic development of areas where land is the most important factor of production. For reasons not discussed here, continued use of this land is often considered as an obvious goal of rural development policy. Consequently, the identification of alternative economic activities which could substitute for land use by agriculture – especially at a regional level – has been a major priority for policy makers and academics. Forestry provides '. . . the greatest alternative use of agricultural land . . . ' (Scottish Development Department, 1987), and increased production could lower the quite considerable costs of timber imports both in a UK and European context. Equally, as a heavily subsidised industry, its expansion would also incur costs: as a result, forestry has occupied a central role in this kind of land-use analysis, conducted not only at the technical and behavioural level of tree planting, maintenance and harvesting at farm level, but also through the application of quantitative techniques such as input-output and cost-benefit analysis at the aggregate or policy level.

The examination of the role of forestry as an alternative land-using rural sector (at both national and regional level) through the use of input-output analysis is not free of data-related and conceptual methodological issues. Most of these issues are connected to key characteristics of forestry: mainly the very long production period which characterises the industry, but also the underlying assumptions (simplifications) of the input-output technique, and the characteristics of forestry data collection and its publication process in the United Kingdom. Our objective in this chapter is to define and analyse these methodological issues. The four main sections deal, respectively, with: a presentation and discussion of conceptual methodological issues in forestry input-output modelling; data-related methodological issues; recreational and other aspects related to forestry; and a discussion of the implications of these issues, with an indication of approaches that could address them.

Conceptual issues

Standard input-output models are based on input-output flows during an annual period. When used to analyse the impact of demand changes, as in multiplier analysis, this approach assumes implicitly that primary and intermediate inputs can be made available and used to produce output mainly for sale (unsold production accruing to the 'stocks' row) within twelve months. While convenient and perhaps appropriate for the bulk of the economy (particularly the growing service sector), this convention can hardly be adopted for forestry, with its production cycles of 40 to 100 years. Similar but much less sizeable problems arise in the construction sector, and even for agricultural livestock production. Clearly, for forestry, much input use results 'only' in additions to growing stock, for which market demand is of an entirely different nature from that for timber. For example, in the 1989 input-output tables for Scotland, more than 99 per cent of the value of input use in the forestry planting sector resulted in additions to growing stocks. All timber production and sales arise from stocks, either at a transitional stage (thinnings) or at final harvest. Again, in the 1989 Scottish tables all raw timber is provided to the Forestry Harvesting Sector by Sales by Final Demand.

Three alternative approaches suggest themselves in terms of applying the input-output method to forestry, given relatively marginal changes in demand for wood output. One is to assume the normal annual framework, in which case growing stocks may be considered sufficient in size and maturity (adequate proportions of trees in the upper ranges of the age distribution) to cope with the acceleration or deceleration in the rate of harvesting implied by the demand change. Planting is in effect a completely separate activity, although its rate may be implicitly determined by that of felling (since, at least in the United Kingdom, the issue of felling licences by the Forestry Authority is conditional on re-stocking). The other extreme is to assume a long enough period (say 50 years for conifer forestry) for the stocks position to become much less 'fixed' and for sufficient planting response to take place to meet the envisaged changes (positive or negative) in output demand. A medium-term (say 10-year) outlook makes the stocks position crucial, without planting being able to affect the potential supply situation. The aggregated treatment of the forestry industry as one sector, as assumed above, rather than its disaggregation into (say) planting and harvesting sectors, gives rise to an important issue for forestry input-output analysis. Forestry-related economic activity reflects the degree of maturity of the domestic forest as well as demand (output market) factors. In an input-output analysis context, this distinctive feature (not encountered in any other sector, at least to such a degree) is very important, because of the snapshot nature of the input-output technique. Looking at the forestry sector as a whole, marginal changes in the balance of industry activities would significantly affect multipliers. New investment effects would arise in the planting sector, and harvesting activity would accrue to the harvesting sector. In input-output tables which do not disaggregate forestry, multiplier linkages relate to the structure of the sector at the time of the construction of the table. As an example, in Rural Scotland, 1989 Type II output multipliers for forestry planting and forestry harvesting are 4.38 and 2.04 respectively. If forestry was treated as one sector, the multiplier would have been 2.81. However, if we could double the gross output of the harvesting sector without changing the output of the planting sector, the aggregated forestry Type II output multiplier would be 2.50. According to McGregor and McNicoll (1992), data limitations preclude a possible solution to this problem, namely the introduction of disaggregated forestry activities in the model.

The appropriate treatment of self-employment income is another important issue for forestry input-output analysis. According to UK employment statistics, a very significant share of the forest industry labour is self-employed. Therefore, their incomes do not appear as 'Wages and Salaries', but in the 'Profits and Other Value Added' row of the input-output table. As a consequence, their incomes are not supposed to create further economic activity (the induced effect), leading to the underestimation of Type II multipliers. However, the low-wage (and low-income) nature of these consumers hardly corresponds to a zero level of marginal propensity to consume. As a possible solution to this problem, we could assume proportionality between wages and salaries per full-time equivalent and self-employment earnings per full-time equivalent, and transfer the appropriate amount to the 'Wages and Salaries' row. In the context of 1989 input-output tables for Rural Scotland, there are considerable implications, especially for Type

II output multipliers, all income coefficients, direct, indirect and induced employment coefficients, and Type II employment multipliers (Table 4.1).

The residence of the forest labour force is another consideration. On a regional scale, inter-regional mobility of workers on individual projects and contracts for considerable periods of time is a problem, in the sense that a significant part of "Wages and Salaries" paid to forest workers, many of whom do not reside locally, may be spent outside the study area. According to various sources (Mather and Murray, 1988; Forestry Commission, 1990), high levels of labour mobility are quite common in forestry. To ignore this issue would lead to the overestimation of regional Type II multipliers, but to make appropriate adjustments in the 'Income from Employment' entry, information about the workers' residence is necessary. In the case of forest workers in 'Rural Scotland', survey information indicated endogeneity of income from employment and household consumption flows (Psaltopoulos and Thomson, 1993).

Lastly, with regard to the forward linkages of the harvesting sector, attention should be given to the substitutability between domestically produced and imported timber used by the processing industry as raw material. There seems no reason to reject perfect substitutability between increases in home-grown and imported timber, at least for lower-quality softwoods in Britain. Higher domestic wood production would thus result in import replacement, assuming no capacity constraints in the wood-processing industry. As a result, there would be an increase in wood-processing multiplier linkages. On the other hand, the absence of an international market for roundwood implies that a possible contraction of domestic forestry could impose a supply constraint, and a subsequent contraction by wood processors for whom import substitution is not feasible. In such a case, the economy-wide forestry-induced losses in output, income, and employment could be significant (Forestry Commission, 1991; McGregor and McNicoll, 1992).

Data-related issues

A common problem with regard to inter-industry transactions of the forestry industry is the fact that they are not separately identified in national input-output tables, which aggregate either Forestry and Fishing (1984 UK input-output tables) or Agriculture, Forestry Harvesting, and Fishing (1985 input-output tables for Ireland: the Irish input-output tables also aggregate Forestry Development, which is establishment, planting, maintenance, with Building and Construction; the total output of the forestry development sector is accounted as Gross Fixed Capital Formation) (Central Statistical Office, 1988, 1992). Therefore, the augmentation of official published tables is often a first research task. Two alternative ways exist for accomplishing this. The appropriate Statistical Office could be approached for the relevant detailed information, which is often

Table 4.1. Output, income and employment multipliers for forestry, rural Scotland, 1989: simple and adjusted for self-employment.

		Forestry Planting	
	Simple(1)	*Adjusted(2)*	*Difference (2-1)*
Output Type I	1.16	1.16	0.00
Output Type II	4.38	6.47	2.09
Direct Income Coefficient	1.49	2.48	0.99
Direct & Indirect Income Coefficient	1.55	2.54	0.99
Income Type I	1.04	1.02	-0.02
Direct, Indirect & Induced Income Coefficient	2.09	3.45	1.36
Income Type II	1.40	1.39	-0.01
Direct Employment Coefficient	84	84	0
Direct & Indirect Employment Coefficient	89	89	0
Employment Type I	1.06	1.06	0
Direct, Indirect & Induced Employment Coefficient	132	160	28
Employment Type II	1.36	1.51	0.15
		Forestry Harvesting	
	Simple(1)	*Adjusted(2)*	*Difference (2-1)*
Output Type I	1.12	1.12	0.00
Output Type II	2.04	2.58	0.54
Direct Income Coefficient	0.38	0.64	0.26
Direct & Indirect Income Coefficient	0.45	0.70	0.25
Income Type I	1.16	1.10	-0.06
Direct, Indirect & Induced Income Coefficient	0.60	0.95	0.35
Income Type II	1.57	1.49	-0.08
Direct Employment Coefficient	47	47	0
Direct & Indirect Employment Coefficient	52	52	0
Employment Type I	1.10	1.10	0
Direct, Indirect & Induced Employment Coefficient	64	71	7
Employment Type II	1.36	1.51	0.15

Source: Authors' calculations; see also Psaltopoulos (1995: 62-66).

collected but not published for lack of completeness (Johns and Leat, 1986; Thomson and Psaltopoulos, 1993). Alternatively, superior information derived from sectoral survey may be used to separate the forestry sector's sales and purchases from the aggregated accounts presented in the original tables (McGregor and McNicoll, 1992).

Another data-related difficulty in forestry input-output modelling is the meaningful estimation of the change in value of standing timber, necessary for the definition of the industry's output which includes timber and other production which is sold, recreation services, and the value of the physical increase in harvesting stocks. In the UK input-output tables, the increase in value of standing timber is included in the 'Stock Change' column (Central Statistical Office, 1968). The Scottish input-output tables distinguish two forestry industry groups, namely Forestry Planting and Maintenance, and Forestry Harvesting, in order to allow for a clear distinction between the two most important employment-creating procedures in the forest cycle. The former sector sells its output (standing timber included) to the 'Stock Change' column, while the latter buys standing timber from the 'Sales by Final Demand' row. However, it is not very clear if the increase in the stocks entry of the planting sector includes the standing timber growth unrelated to new planting, as well as appreciation. The limited availability of information on private forestry is a further major problem in estimation of the increase in the value of standing timber. Estimates can be made either by carrying out a comprehensive survey of private forestry on a frequent basis (an expensive solution), or by assuming similarity to Forestry Commission output patterns (Industry Department for Scotland, 1984). There is often limited information on private-sector timber sales, and recreation-related income. The former may be estimated using unit values for Forestry Commission sales, and the latter by assuming proportionality to the area of productive woodland.

Lack of information on private forestry wages and salaries is another data-related problem. So far (e.g. 1989 Scottish input-output tables), estimates have been made by assuming that the functional employment breakdown for Forestry Commission and private woodlands employees holds also for wages and salaries.

On a regional (e.g. Forest District or Local Authority District) level, it is extremely difficult to acquire any information on the private sector necessary for exercises such as the improvement of mechanically derived regional technical coefficients.

Another important data-related problem is related to the availability of employment and self-employment data, both at national and regional level. On a national level, 'employees in employment' as well as 'self-employment' data are not available on a forest planting-harvesting breakdown. Also, self-employment statistics are subject to relatively high sampling error (10 per cent sample), are aggregated together in the composite forestry and fishing classification, and provide no information about part-time self-employment. Information provided by the Forestry Commission tri-annual forest employment survey could be the solution to all these problems. However, this survey provides information on employment only in the form of whole-job equivalents by forest operation (Forestry Commission, 1990). Because, as discussed in the previous section, there is no distinction between employees and self-employed, this is of little concrete value.

At a regional level, problems relating to employment data-related become even more difficult to handle. In the United Kingdom, Forestry Commission survey data is only available on a country (England, Scotland, Wales) basis, and although 'employees in employment' and 'self-employment' data are broken down in detail geographically, they are certainly not free from significant problems, as mentioned above, and, more importantly, cannot distinguish between forest planting and harvesting.

Recreational and other aspects

Much recent interest in forestry economics has focused on non-timber aspects of the industry, especially non-market commodities such as 'free' recreation, nature conservation, water catchment and carbon-fixing. Input-output analysis, which explicitly models physical and financial flows within the *market* economy, is not well suited as a technique to address these problems. However, certain aspects can be at least addressed, specifically recreational use of forests, especially where this involves payment for services by forest users or visitors, either directly to forest owners, or indirectly for transport, food, accommodation, and other services.

For example, travel-cost estimates of forest-user expenditure associated with using the forest resource within a regional economy can be allocated as part of final demand to sectors defined in an input-output model for that economy (Jones and Stokes Associates, 1987). Output, income, and employment impacts resulting from forest-user spending can therefore be estimated. There are difficulties with identifying only user expenditures strictly associated with use of the forest, when other sites are also visited, and at a regional level, expenditure information needs to be available for the economy under study, in order to account for regional leakages.

Conclusions

The application of input-output analysis to forestry has been shown to throw up a number of conceptual and data-related problems, some of which are difficult to resolve given the issues of economic analysis usually addressed, and the quality and structure of the data available. For certain questions, the input-output method will not be appropriate, and should not be forcibly applied. For others, particularly in respect of the role which forestry, alongside agriculture, does and could play in the general rural development of a region or country, the approach seems worthwhile if some of the problems addressed in this chapter can be overcome. However, it is clear that care is needed in applying the method to the forestry sector of the rural economy, and perhaps even greater care in interpreting both questions and answers between analysts and their audiences.

References

Central Statistical Office (1968). *National accounts statistics*, London: HMSO.

Central Statistical Office (1988). *Input-output tables for the UK 1984*, London: HMSO.

Central Statistical Office (1992). *Input-output tables for 1985*, Dublin: Central Statistical Office.

Forestry Commission (1990). *Forest employment survey, 1988-1989*, Edinburgh: Forestry Commission.

Forestry Commission (1991). *Forestry expansion: a study of technical, economic and ecological factors*, Edinburgh: Forestry Commission.

Gardiner, J.J., Thomson, K.J., Davis, J., Psaltopoulos, D., Ni Dhubhain, A., Hutchinson, G.W., and Chilton, S. (1994). *The socio-economic impact of afforestation in rural development,* CAMAR Contract no. 8001-CT90-0008, Report to the European Commission.

Industry Department for Scotland (1984). *Scottish input-output tables for 1979*, Vols. 1-5, Edinburgh: HMSO.

Jensen, R.C., Mandeville, T.D. and Karunaratne, N.D. (1979). *Regional economic planning: generation of regional input-output analysis*, London: Croom Helm.

Johns, P.M. and Leat, P.M.K. (1986). *An approach to regional economic modelling: the case of Grampian*. Economic Report No. 144, North of Scotland College of Agriculture, Aberdeen: NoSCA.

Jones and Stokes Associates (1987). *Juneau area sport fishing economic study*, Sacramento: Jones and Stokes Associates.

Leontief, W.W. (1951). *Studies in the structure of the American economy 1919-1939*, Cambridge: Cambridge University Press.

Mather, A.S. and Murray, N. (1988). The dynamics of rural land-use change: the case of private sector afforestation in Scotland, *Land Use Policy,* 5, 103-120.

McGregor, P.G. and McNicoll, I.H. (1992). The impact of forestry on output in the UK and its member countries, *Regional Studies*, 26, 69-79.

Miernyk, W.K. (1969). *Elements in input-output analysis*, New York: Random House.

Psaltopoulos, D. (1995). Input-output analysis of Scottish forestry strategies, unpublished PhD Thesis, Aberdeen: University of Aberdeen.

Psaltopoulos, D. and Thomson, K.J. (1993). Input-output evaluation of rural development: a forestry-centred application, *Journal of Rural Studies,* 9, 351-358.

Scottish Development Department, (1987). *Land use summary sheet – agriculture*, Edinburgh: Scottish Development Department.

Thomson, K.J. and Psaltopoulos, D. (1993). The impact of forestry on output and employment in Scotland and Ireland. In: *Forestry and rural development in industrialised countries: where are we going?,* IUFRO Symposium Proceedings, Fredericton: Forestry Canada.

Chapter five:
Income Distribution and the Structure of Production: Insights from a SAM-based Model of UK Agriculture

Debbie Roberts and Noel Russell
Centre for Agricultural, Food and Resource Economics, School of Economic Studies, The University of Manchester

Introduction

Of all the justifications for government support of the agricultural sector, concern with the absolute and relative level of farm incomes has been given the greatest weight in most developed countries. Indeed, the need to maintain farm incomes remained unquestioned throughout the recent CAP reforms and GATT negotiations. It is thus paradoxical that the method of farm support most widely employed in the same developed economies, price support, is not only extremely inefficient in achieving this objective but, by its very nature, necessarily increases the inequality with which incomes are distributed within the farm sector. It is therefore not surprising that the move towards a more 'decoupled' system of farm support under the 1992 McSharry reforms has been greeted with much enthusiasm.

The direct income payments agreed as part of the CAP reform package are, at present, mostly tied to an input in the production process. For example, the arable acreage payments are contingent upon setting aside a certain area of land whilst sheep and beef premiums are linked to numbers of livestock and stocking rates. Thus the link between support and production levels has yet to be completely severed. However, in future it is highly likely that such payments will become more 'purely' decoupled, allowing the benefits to be targeted to certain types of farms or particular categories of farm households in the economy. They offer the opportunity for agricultural policy to become, as Fennell (1993) suggests, 'more closely attuned to the needs of those who work in the sector and the rural space in which it functions'. Just as agricultural price support has been, in part, justified by the positive multiplier effects production on farms has on those employed in the upstream and downstream sectors, direct income payments to farm households could, in part, be justified by the positive multiplier effects private farm household expenditure has on the rural economy in which they are located. However, to target direct income payments effectively and efficiently it is necessary to know something of the nature of income generation and income distribution in an economy.

The level of income received by a particular type of household within an economy is a function of, amongst other factors,

(a) the linkages between households and production sectors via the value-added payments made in return for household services,
(b) the extent and pattern of inter-industry linkages (as captured by traditional input-output models)
and (c) the pattern of household final consumption, and inter-household/inter-institutional transfers.

These three features will, likewise, determine the 'multiplier' effects arising from a direct income payment to a particular farm household category. This chapter argues that, by taking all of the above into account, a Social Accounting Matrix (SAM) based model can elucidate the nature of income generation and distribution and thus aid the targeting issue. In other words, a SAM-based model can be used to identify 'key institutions' , analogous to key sectors in input-output analysis, and thus indicate the types of farmers towards which decoupled income payments would be most efficiently directed.

The chapter is structured as follows. In the following section, the nature of a modified SAM Leontief model is briefly rehearsed, concentrating on the way in which it elucidates the interdependency between the structure of production and distribution of household incomes. Further, it is described how the re-distributional impacts of different types of economic policy decisions can be determined, concentrating, in particular, on the identification of relative gainers/losers from the switch from coupled to more decoupled farm income support. The third section reports some empirical findings based on a SAM which emphasises the role of the agricultural sector in the UK economy. The results from the UK SAM can, at best, only illustrate the potential of the methodology in identifying the most efficient targeting of direct income payments. The final section concludes by discussing how the modelling framework, given an appropriate rural or regional database, can be used to investigate the type of agricultural policy issues outlined above, and discusses particular aspects of rural economies that will influence whether the benefits of decoupled farm income support are retained within the region/rurality or are leaked to other areas of the economy.

Methodology

To investigate the nature of income generation and the pattern of income distribution between houschold typcs, a modelling framework is required that explicitly recognises the full circular flow of income around the economy. This, in turn, requires a benchmark data set sufficiently disaggregated to reflect, at a minimum, the following three features: firstly, that different production sectors have different patterns of factor payments; secondly, that each type of factor payment (for example wages and salaries, rental income, pensions, etc.) are distributed between household categories in a different fashion; and thirdly, that the data framework should be sufficiently disaggregated to reflect the diverse patterns of expenditure of different types of households. A Leontief model based on a SAM data framework satisfies these methodological requirements and solves the mutual relationship between the sectoral distribution of output and the distribution between households simultaneously within one consistent system.

The basic nature of the SAM-based Leontief model has often been recapitulated (Pyatt and Roe, 1977; Pyatt and Round, 1979; Adelman and Robinson, 1986). By specifying that at least one of the accounts within the SAM is exogenously determined (typically, the government[1] and/or 'rest of world' account), one can derive a coefficients matrix, **B***, showing the pattern of direct linkages between all of the remaining (endogenously determined) accounts in the model. For example, with only the production, factor and household accounts treated as endogenous,

$$\mathbf{B}^* = \begin{bmatrix} \mathbf{A} & \mathbf{0} & \mathbf{C} \\ \mathbf{V} & \mathbf{0} & \mathbf{0} \\ \mathbf{0} & \mathbf{D} & \mathbf{T} \end{bmatrix} \qquad [5.1]$$

where, with m production activities, n factors, and k endogenous institutions, **B*** is the complete SAM coefficients matrix $(m+n+k,m+n+k)$; **A** is a matrix of input-output coefficients (m,m); **V**, a matrix of value-added coefficients (n,m); **C**, a matrix of current account expenditure coefficients (m,k); **D**, a matrix of coefficients reflecting how factor income is distributed between institutional categories (k,n) and **T**, a matrix of transfer coefficients between institutional categories (k,k).

The accounting balances of the benchmark data set can be written:

1 The endogenous treatment of the government sector is, of course, required if the intention is to use the model to investigate the impact of policy changes/contrast the effects of coupled versus decoupled farm income support.

$$\mathbf{B}^* \begin{bmatrix} \mathbf{p} \\ \mathbf{v} \\ \mathbf{z} \end{bmatrix} = \begin{bmatrix} \mathbf{e}_p \\ \mathbf{e}_v \\ \mathbf{e}_z \end{bmatrix} \qquad [5.2]$$

where $\mathbf{p}$ is a vector of gross output levels of the production sectors $(m,1)$; $\mathbf{v}$, a vector of total value-added payments by factor type $(n,1)$; $\mathbf{z}$, a vector of total household incomes $(k,1)$; and $\mathbf{e}_p$, $\mathbf{e}_v$, $\mathbf{e}_z$ vectors of injections from the exogenous accounts to the production, factor and household accounts respectively.

Assuming that the elements in $\mathbf{B}^*$ remain constant, at the margin, the basic SAM Leontief model can be written simply as:

$$\begin{bmatrix} \mathbf{p} \\ \mathbf{v} \\ \mathbf{z} \end{bmatrix} = (\mathbf{I} - \mathbf{B}^*)^{-1} \begin{bmatrix} \mathbf{e}_p \\ \mathbf{e}_v \\ \mathbf{e}_z \end{bmatrix} = \mathbf{M} \begin{bmatrix} \mathbf{e}_p \\ \mathbf{e}_v \\ \mathbf{e}_z \end{bmatrix} \qquad [5.3]$$

The composition of the multiplier matrix, $\mathbf{M}$, is expressed in its partitioned form as:

$$\mathbf{M} = \begin{bmatrix} \mathbf{Q+QCZDVQ} & \mathbf{QCZD} & \mathbf{QCZ} \\ \mathbf{VQ+VQCZDVQ} & \mathbf{I+VQCZD} & \mathbf{VQCZ} \\ \mathbf{ZDVQ} & \mathbf{ZD} & \mathbf{Z} \end{bmatrix} \qquad [5.4]$$

where $\mathbf{Q} = (\mathbf{I} - \mathbf{A})^{-1}$
and $\mathbf{Z} = [(\mathbf{I} - \mathbf{T}) - \mathbf{DVQC}]^{-1}$

It is the elements within the final row of this partitioned matrix that indicate the direct, indirect and induced impact on the incomes of each household group arising from an injection from an exogenous account. These are the multipliers that form the focus of this chapter.

Explicitly, sub-matrix $\mathbf{M}_{31}$, (= $\mathbf{ZDVQ}$) shows the impact on household incomes of an exogenous increase in demand for output from the production sectors. The composition of these multipliers indicates that the increased income generated by a production-led injection depends on the pattern of interconnections between all of the various types of endogenous accounts in the economy. Sub-matrix $\mathbf{M}_{32}$ (= $\mathbf{ZD}$) shows the income generated from exogenous injections to the factor accounts. Again, these multiplier effects depend on features of the economy other than simply the direct distribution of factor earnings between household groups. Finally, $\mathbf{M}_{33}$ (= $\mathbf{Z}$) indicates the total multiplier impact of exogenous injections to household incomes after the impact has worked its way around the whole economy. Thus interconnections between the current accounts of the economic system effect the magnitude and distribution of benefits between household groups arising from all three types of exogenous injections.

Although the derivation of the multipliers described above would go some way towards identifying the relative efficiency of different types of income-enhancing policies, one might also be concerned with the redistributional impact of a change in economic policy. Roland-Holst and Sancho (1992) restate a formal model of relative income determination as it applies to a SAM Leontief model so as to derive a matrix, **R**(x), of redistribution effects capable of indicating how the structure of the economic system influences the relative gainers/losers from a change in the nature of policy.

Following on from above, the redistribution model takes the form

$$\begin{aligned} dy &= [\mathbf{i}'\,\mathbf{M}\,\mathbf{x}]^{-1}\,[\mathrm{I}\text{-}[\mathbf{i}'\,\mathbf{M}\,\mathbf{x}]^{-1}\,(\mathbf{M}\,\mathbf{x})\mathbf{i}']\mathbf{M}\,dx \\ &= \frac{1}{\mathbf{i}'\,\mathbf{Y}}\left[\mathbf{I} - \frac{\mathbf{Y}}{\mathbf{i}'\,\mathbf{Y}}\mathbf{i}'\right]\mathbf{M}\,dx \\ &= \mathbf{R}(\mathrm{x})\,dx \end{aligned} \qquad [5.5]$$

where **i**' denotes a unit row vector, **Y** the $(m+n+k,1)$ column vector of the initial total incomes of the endogenous accounts and x the $(m+n+k,1)$ vector of injections from the exogenous accounts.

R(x) is the redistribution matrix indicating how relative incomes are affected by a change in exogenous injections, dx. Each individual element of **R**(x), R_{ij}, is given by:

$$R_{ij} = \frac{1}{\mathbf{i}'\,\mathbf{Y}}\left[M_{ij} - \frac{Y_i}{\mathbf{i}'\,\mathbf{Y}}\mathbf{i}'\mathbf{M}_{\bullet j}\right] \qquad [5.6]$$

where $\mathbf{M}_{\bullet j}$ is the *j*th column of **M**. $\mathbf{i}'\mathbf{M}_{\bullet j}$ is thus the *j*th column sum of **M**. R_{ij} will be positive only if

$$\frac{M_{ij}}{\mathbf{i}'\mathbf{M}_{\bullet j}} > - \frac{Y_i}{\mathbf{i}'\,\mathbf{Y}} \qquad [5.7]$$

That is, any injection into account *i* will be beneficial to account *i* only if its share of the multiplier effect exceeds its share of initial nominal income. Alternatively, R_{ij} will be negative if

$$\frac{M_{ij}}{\mathbf{i}'\mathbf{M}_{\bullet j}} < - \frac{Y_i}{\mathbf{i}'\,\mathbf{Y}} \qquad [5.8]$$

which indicates that account i will (relatively) lose out in terms of income distribution because its share of the multiplier gains from the injection is less than its share of initial nominal income. The following section contrasts the re-distributional effects on household incomes of injections to the three different types of endogenous accounts in the SAM model, revealing the way in which relative incomes are affected by different types of exogenous forces.

Since, like its input-output counterpart, the basic SAM Leontief model contains no policy variables, in order to carry out any sort of policy analysis, one has to translate or reinterpret the impact of specific policy instruments as injections to the system from the exogenous accounts. Although the level of disaggregation and the choice of classification system used in the SAM does play a large part in determining the quality of policy analysis that is possible, this translation is not always easy. For the purpose of this chapter, the different effects arising from the switch from coupled to decoupled farm income support are captured by modelling coupled farm support as exogenous injections through the production accounts, decoupled farm income support as injections to either the factor or household accounts. The argument in the case of the former is that the benefits to farm households (and, through multiplier effects, other households in the economy) from coupled farm policies originate in the form of demand for agricultural output. On the other hand, direct injections to household incomes bear closest resemblance to what most would consider to be meant by decoupled income support although, importantly, the income supplements are not production neutral within the model in that extra activity may be stimulated indirectly as the impact works its way around the system. The effects arising from injections to the factor accounts are also explored in the empirical analysis that follows primarily because, given the structure and classification of the accounts in the SAM, these could most closely be interpreted as injections to farm household incomes.

Empirical analysis

As mentioned above, the SAM on which the following analysis is based was constructed with the intention of emphasising agriculture's role within the UK economy. This is reflected in the choice of classification system and level of disaggregation of the accounts in the matrix, particularly in the case of the production accounts.[2]

The issue of how to disaggregate private households within the SAM framework was resolved by the availability of data which needed to be fully articulated – showing both the origin and destination of each transaction. Despite the arguments against using the size distribution of income as the basis of

2 Twelve out of the total 37 commodity accounts in the SAM are farm outputs whilst the industry accounts distinguish twenty-three different farm types. Further, the major (direct) upstream and downstream industries in the food chain are separately identified.

classification (see Pyatt and Thorebecke, 1976; Kennes and van Veen, 1985; and Keuning and Ruijter, 1988[3]), private households in the UK were split into income quintiles. This at least made observations on the (in)equality of distribution of personal income in the economy easy to assess.

Importantly, before computing the multiplier matrix, **M**, from the SAM, the household average propensities to consume elicited from the SAM coefficients matrix (that is sub-matrix **C**, in **B*** above) were replaced with marginal propensities to consume calculated using extraneous estimates of the income elasticities of demand for each of the commodities in the model. Thus, by following the example of Pyatt and Round (1979), the most limiting assumption of the SAM Leontief model was removed (that every household's marginal response to an increase in income is equal to their average response) and so-called fixed price as opposed to accounting multipliers were formed. Full documentation of the SAM and sources of estimates of income elasticities are given in Roberts (1992).

Results

Table 5.1 presents the nominal multiplier effects on the income of the five household categories (from Q1, the poorest quintile, through to the richest, Q5) arising from a unit change in (exogenous) demand for ten different types of commodity output, and thus focuses on the relationship between income levels and the value of sales from the production sectors in the economy. In the first instance, agricultural commodities are aggregated into a single account to facilitate comparisons with other broad types of commodities in the economy. The multipliers in Table 5.1 correspond to sub-matrix $\mathbf{M}_{31}$, described in the section above and, with government treated as exogenous, arise from what can broadly be interpreted as coupled government support policies.

The column totals in Table 5.1 suggest that public administration, service and distribution sectors hold the greatest potential for increasing total household income in the economy as might have been anticipated given that these industries are all labour intensive and thus have a high proportion of value-added contributing directly to household incomes in the form of wages and salaries. An exogenous increase in the demand for output from the agricultural sector on the other hand has a fairly low multiplier impact on total household incomes, with only £693 thousand generated in private incomes from a £1 million injection: compare this to the £1.227 million generated through public administration. Further, Table 5.1 suggests that all of the exogenous injections via the production accounts boost the absolute income levels of the richest households far more than the lower income groups. Table 5.2 investigates this point further by indicating

3 These authors argue that classification of household groups should be based on criteria such as geographical location, sociological considerations and/or wealth rather than income levels since households grouped in accordance with the latter do not necessarily exhibit homogeneous behaviour and are not the usual target for policy reform.

the manner in which income is redistributed between households as a result of an increase in demand for each type of commodity.

Table 5.1. Nominal multiplier effects on household incomes of exogenous shocks to production accounts.

	Agric.	*Energy*	*Gen. Manuf.*	*Agric. Inputs*	*Process. Foods*	*Construc-tion*
Q1	0.009	0.011	0.005	0.007	0.007	0.009
Q2	0.049	0.056	0.040	0.044	0.044	0.061
Q3	0.120	0.124	0.116	0.119	0.117	0.162
Q4	0.185	0.182	0.190	0.190	0.186	0.257
Q5	0.331	0.333	0.334	0.337	0.330	0.457
Column Total	0.693	0.706	0.686	0.697	0.682	0.946

	Distrib-ution	*Trans-port*	*Services*	*Public Admin.*	*Average*	*Row Total*
Q1	0.008	0.007	0.010	0.011	0.008	0.084
Q2	0.059	0.048	0.063	0.075	0.054	0.538
Q3	0.164	0.132	0.168	0.209	0.143	1.429
Q4	0.264	0.211	0.268	0.337	0.227	2.268
Q5	0.467	0.374	0.476	0.596	0.403	4.034
Column Total	0.961	0.773	0.984	1.227	0.835	8.354

Having assessed how agriculture compares on aggregate with other types of production sectors, Tables 5.3 and 5.4 replicate the preceding analysis but, in these cases, concentrate exclusively on the farm sector by considering exogenous increases in demand for specific agricultural commodities. Again, Table 5.3 indicates significant differences between the income-generating potential of 'coupled' farm policies depending on which specific commodity is supported. In general, government demand for livestock products would seem to have the most beneficial impact on total private household incomes but, from Table 5.4, the majority of this benefit accrues to the richest household category. The magnitude and pattern of effects recorded in Tables 5.3 and 5.4 is determined by the mechanisms explicitly incorporated in the SAM Leontief model, that is interactions within and between the production, factor and household accounts in the model. These interactions suggest that the distributive effects of coupled policies (which by their very nature favour larger, in general richer, farmers more than smaller producers) are compounded by the nature of income generation in the economy since, taking into account the indirect and induced effects of the injection, support given in the form of increased demand for commodity output increases the overall inequality of income distribution in the economy.

Table 5.2. Redistributional impact of exogenous shocks to production accounts.

	Agric.	*Energy*	*Gen. Manuf.*	*Agric. Inputs*	*Process. Foods*	*Construc-tion*
Q1	-0.028	-0.026	-0.031	-0.030	-0.030	-0.041
Q2	-0.024	-0.018	-0.032	-0.029	-0.028	-0.039
Q3	0.001	0.020	-0.002	-0.001	-0.001	-0.001
Q4	0.013	0.007	0.019	0.017	0.016	0.022
Q5	0.039	0.036	0.045	0.044	0.042	0.058
Column Sum	0.000	0.000	0.000	0.000	0.000	0.000
Change	0.052	0.044	0.065	0.060	0.059	0.081

	Distrib-ution	*Trans-port*	*Services*	*Public Admin.*	*Average*
Q1	-0.043	-0.034	-0.043	-0.054	-0.036
Q2	-0.042	-0.033	-0.041	-0.054	-0.034
Q3	-0.002	-0.001	-0.001	-0.002	-0.001
Q4	0.025	0.019	0.023	0.032	0.020
Q5	0.062	0.049	0.061	0.079	0.052
Column Sum	0.000	0.000	0.000	0.000	0.000
Change	0.087	0.068	0.085	0.110	0.072

Table 5.3. Nominal multiplier effects on household incomes of exogenous increase in demand for agricultural commodities.

	Cereals	*Potato*	*Other crops*	*Cattle*	*Sheep*	*Wool*	*Milk*
Q1	0.009	0.009	0.006	0.011	0.012	0.003	0.010
Q2	0.052	0.053	0.033	0.064	0.066	0.015	0.060
Q3	0.124	0.127	0.080	0.156	0.162	0.036	0.150
Q4	0.190	0.195	0.122	0.242	0.251	0.055	0.232
Q5	0.341	0.350	0.220	0.433	0.449	0.098	0.415
Column Total	0.716	0.735	0.460	0.905	0.939	0.206	0.867

	Pigs	*Eggs*	*Poultry*	*Other Lvstk*	*Misc. Agric*	*Avge*	*Row Total*
Q1	0.009	0.009	0.009	0.011	0.011	0.009	0.110
Q2	0.051	0.049	0.050	0.063	0.062	0.052	0.619
Q3	0.125	0.120	0.123	0.154	0.151	0.126	1.509
Q4	0.192	0.185	0.189	0.236	0.233	0.193	2.322
Q5	0.344	0.332	0.339	0.423	0.418	0.347	4.162
Column Total	0.721	0.696	0.711	0.888	0.876		

Table 5.4. Redistributional effects of an exogenous increase in demand for agricultural commodities.

	Cereals	*Potato*	*Other crops*	*Cattle*	*Sheep*	*Wool*	*Milk*
Q1	-0.029	-0.030	-0.019	-0.037	-0.038	-0.008	-0.036
Q2	-0.023	-0.024	-0.016	-0.032	-0.033	-0.007	-0.031
Q3	0.001	0.000	0.001	0.001	0.000	0.000	0.000
Q4	0.012	0.013	0.008	0.017	0.017	0.004	0.017
Q5	0.039	0.041	0.026	0.051	0.053	0.012	0.050
Column							
Sum	0.000	0.000	0.000	0.000	0.000	0.000	0.000
Change	0.052	0.054	0.035	0.069	0.071	0.015	0.067

	Pigs	*Eggs*	*Poultry*	*Other Lvstk*	*Misc. Agric*	*Avge*
Q1	-0.029	-0.028	-0.029	-0.036	-0.036	-0.029
Q2	-0.024	-0.024	-0.024	-0.030	-0.030	-0.025
Q3	0.001	0.001	0.001	0.001	0.001	0.001
Q4	0.013	0.012	0.013	0.016	0.016	0.013
Q5	0.040	0.039	0.040	0.049	0.049	0.041
Column						
Sum	0.000	0.000	0.000	0.000	0.000	0.000
Change	0.053	0.052	0.053	0.066	0.066	0.054

Tables 5.5 and 5.6 indicate the multiplier and redistributional effects that arise from exogenous stimuli to the household rather than production accounts. Such impacts correspond to what most would consider to be a decoupled policy instrument – direct income supplements from government to a household category – although, in terms of their composition, the multipliers in Table 5.5 relate to submatrix $\mathbf{M}_{33}$, described in the section above.

Table 5.5. Nominal multiplier effects arising from an injection direct to household incomes.

	Q1	*Q2*	*Quintile Q3*	*Q4*	*Q5*	*Avge*	*Total*
Q1	1.007	0.007	0.006	0.005	0.005	0.206	1.030
Q2	0.044	1.041	0.036	0.033	0.030	0.237	1.184
Q3	0.120	0.112	1.098	0.090	0.082	0.300	1.502
Q4	0.193	0.180	0.157	1.144	0.131	0.361	1.805
Q5	0.341	0.318	0.279	0.254	1.232	0.485	2.425
Column							
Total	1.705	1.658	1.576	1.526	1.480	1.589	7.945

Table 5.6. Re-distributive effects of an injection direct to household incomes.

			Quintile		
	Q1	*Q2*	*Q3*	*Q4*	*Q5*
Q1	0.916	-0.082	-0.078	-0.076	-0.073
Q2	-0.135	0.867	-0.130	-0.127	-0.125
Q3	-0.173	-0.173	0.827	-0.173	-0.173
Q4	-0.231	-0.232	-0.234	0.765	-0.236
Q5	-0.377	-0.380	-0.385	-0.388	0.608
Column					
Sum	0.000	0.000	0.000	0.000	0.000
Change	0.916	0.867	0.827	0.765	0.608

Whilst the poorest income quintile appears, from Table 5.5, to have the greatest potential for generating private household income (a £l million injection to the 20% of households with the lowest incomes in the base year of the model generating, in total, £l.7 million once the impact has toured through the whole economy), these households are unable to retain much of the benefit for themselves. The total 'own' multiplier for the lowest quintile group was found to be only £1.006 million, as compared to £l.232 million for the richest group of households. These figures are of relevance when considering the possible knock-on effects arising from the switch to decoupled support. If, for example, farm household behaviour most closely corresponds to that of the lowest income group of households in the model (both in terms of income and expenditure), the results suggest that direct income payments to farmers will (a) generate very little benefit to farm incomes above and beyond the value of the initial injection but (b), through farm household 'ways of spending', be of considerable benefit to higher income groups. Alternatively, if farm household behaviour more closely corresponds to that of the richest quintile group, the knock-on or multiplier benefits of decoupled farm income support to non-farm households will be of limited magnitude.

The inability to identify farm households within the model obviously limits the value of the preceding analysis in terms of identifying the consequence of a change in agricultural policy. However, the factor accounts of the model, distinguishing between various different forms of income, allow closer analysis of how the switch from coupled to more decoupled farm income support might affect income generation and distribution in the economy. Table 5.7 compares the nominal multiplier effects arising from injections to two of the factor accounts which distribute income to households in the form of (a) wages and salaries and (b) self-employment income. Definitions of each of these categories are given in the Appendix to this chapter. Since income from farming falls within the second of these categories, one can interpret an injection into this account as a decoupled farm income policy and thus compare the multiplier and distributional effects from such an injection with those arising from coupled policies (Tables 5.1-5.4).

Table 5.7. Nominal multiplier effects from an injection to the factor accounts.

	Wages & Salaries	*Self-employment Income*
Lowest Quartile	0.008	0.018
Second Quartile	0.078	0.127
Third Quartile	0.254	0.261
Fourth Quartile	0.430	0.356
Highest Quartile	0.748	0.763
Column Total	1.518	1.526

Table 5.8. Redistribution effects from an injection to the factor accounts.

	Wages & Salaries	*Self-employment Income*
Lowest Quartile	-0.0728	-0.0628
Second Quartile	-0.0818	-0.0335
Third Quartile	-0.0070	-0.0016
Fourth Quartile	0.0532	-0.0225
Highest Quartile	0.1085	0.1203
Column Sum	0.0000	0.0000
Change	0.1617	0.1203

The magnitude of total private income generated from injections to the factor accounts is larger than that created from injections to the production accounts but similar to the income arising from direct exogenous shocks to the household accounts as would be anticipated.[4] Although both types of factor injections generate total income of a similar magnitude, it is interesting that injections to self-employed income create slightly more benefit to the whole of the private sector than injections to paid employees. Again this can be explained by a series of interconnections between the endogenous accounts in the model, in the first instance, by the distribution of self-employed income between income quintiles. More startling are the redistributional consequences of injections to the factor accounts. As shown in Table 5.8, whilst all the households benefit in nominal terms, the relative gainers and losers depend on which type of factor is stimulated. An exogenous boost to self-employment income which, it has been argued, most closely replicates a switch to decoupled farm income support, increases the relative income of the richest household groups whilst in the case of boosting the level of wages and salaries, the top two income groups benefit at the

4 There is not a one-to-one mapping between total value-added payments from production and total factor earnings of households (since, for example, the government receives a portion of factor payments in the form of employer contributions to social security, etc.). Thus the multiplier effects created by injections to the two sets of accounts are not strictly comparable.

same time widening the gap between their income level and the income of all other households in the economy.

Conclusions

The analysis in this study illustrates the potential usefulness of a SAM-based modelling system for investigating the distributional impacts of changes in agricultural policy payments. The limited policy simulations which we have undertaken have clearly shown that increases in household income are not most efficiently generated by injections through the agricultural production accounts. Public administration, service and distribution sectors, being more labour intensive, have much higher income multipliers. For all these sectors including agriculture, the increase in income is greatest for the richest households. However, agriculture is less regressive in this regard. Within agriculture, livestock production has a higher total impact than crop production, but this is concentrated in the higher income households.

Decoupled policies which provide direct supplements to household income, have more progressive effects especially when the initial stimulus is channelled through the lower income households. However, while the total income boost is between 50 per cent and 70 per cent greater than the initial stimulus, no more than 7 per cent of this accrues to the lowest income households. If farm household behaviour corresponds most closely to this group, the results suggest that direct income payments to farmers have a considerable knock-on effect on non-farm income, though little more than the initial injection accrues to farm households. A decoupled policy which provides a direct boost to factor incomes has an effect on total incomes of similar magnitude though the effect is focused to some extent on higher income households.

These results are presented only as an indication of how this modelling framework might be used. We are well aware of the limitations of the model. In particular we are aware of deficiencies in the data framework in its current implementation. This has meant that while disaggregated production accounts allowed some detailed analysis of coupled policies, similarly disaggregated household accounts did not allow corresponding analyses for decoupled policies, and we are not, therefore, in a position to explore the implications of policy changes for the distribution of income between different categories of farmer and rural households.

Further development would also indicate the desirability of a regionally disaggregated set of accounts. Bemat and Johnson (1991) used a model based on this type of data system to explore differences in response to changes in coupled policies in the US. Their analyses indicated a broadly similar pattern of responses across 12 regions, with largest income increases accruing to households in the middle to highest income groups.

In all modelling efforts in this framework, the inflexibility of Leontief-type relationships is always a source of dissatisfaction, only partly overcome by incorporating some marginal response coefficients in the matrix. A more satisfactory solution, exemplified by a recent study by Kilkenny (1993), is to use the SAM database as a means to calibrate a Computable General Equilibrium model. A bi-regional model, focusing on urban and rural regions, was used to examine the impact of eliminating coupled and decoupled farm support payments in the US. By capturing the potential impact of regional product differentiation and market segmentation, of regional diversification in the ownership of rural resources, and of other leakages outside the rural region, the study showed a net overall gain from subsidy elimination but significant reductions in income and employment in agriculture and related industries and in rural areas in general. The flexibility afforded by this approach has much to commend it.

References

Adelman, I. and Robinson, S. (1986). US agriculture in a general equilibrium framework: analysis with a social accounting matrix, *American Journal of Agricultural Economics*, 68, 1197-1206.

Bemat, A.G. and Johnson, T.G. (1991). Distributional effects of household linkages, *American Journal of Agricultural Economics*, 73, 326-333.

Fennell, R. (1993). *The CAP: unfinished business, unresolved problems*, paper presented to the 7th EAAE Congress, Stresa, Italy, September 1993.

Kennes, W. and van Veen, W. (1985). *Social accounting matrices: theoretical structure and practical difficulties with reference to the Bangladesh agricultural model*, Staff Working Paper SOW-82-18R, Centre for World Food Studies: Amsterdam.

Keuning, S. and Ruijter, W. (1988). Guidelines to the construction of a social accounting matrix, *Review of Income and Wealth*, 34, 71-101.

Kilkenny, M. (1993). Rural/urban effects of terminating farm subsidies, *American Journal of Agricultural Economics*, 75, 968-980.

Pyatt, G. and Roe, A.N. (1977). *Social accounting for development planning with special reference to Sri Lanka*, Cambridge: Cambridge University Press.

Pyatt, G. and Round, J.I. (1979). Accounting and fixed price multipliers in a social accounting framework, *Economic Journal*, 89, 850-873.

Pyatt, G. and Thorebecke, E. (1976). *Planning: techniques for a better future*, Geneva: International Labour Office.

Roberts, D. (1992). UK agriculture in the wider economy: an analysis using a social accounting matrix, unpublished PhD Thesis, Department of Agricultural Economics (University of Manchester).

Roland-Holst, D.W. and Sancho, S. (1992). Relative income determination in the united states: a social accounting perspective, *Review of Income and Wealth*, 38, 311-327.

Appendix

The choice of forms of income used in the SAM was determined mainly by the types of income of the personal, general government and corporate sectors delineated in the National Accounts. However, the categories also had to be complementary to the alternative sources of income recorded in the Department of Employment's Family Expenditure Survey (FES) as this source was used to disaggregate the personal sector into household groups. Accordingly, the two forms of income discussed in the empirical results section of the text are defined as follows:

Wages and salaries

The cash earnings of civilian wage earners and salaried employees, before the deduction of income tax, surtax or insurance contributions. The figures include payments in kind and exclude expenses of employment.

Self-employment income

The income of persons from unincorporated businesses, mainly farmers, professionals who do not receive salaries, shopkeepers and other sole traders and partnerships. In conjunction with the requirements of the SAM, the entries in this account exclude business operating expenses but are calculated prior to the deduction of income tax and interest payments.

Chapter six:
Structural Changes in Final Demands and Productivity Growth

Scott McDonald
Department of Economics, The University of Sheffield

Introduction

Since the pioneering work of Solow (1957), estimates of aggregate (economy-wide) and industry[1] total factor productivity growth have appeared regularly in the literature. But, despite the now considerable body of research on productivity growth rates, relatively little attention has been devoted to the analysis of the impact of changes in output structure upon the productivity growth performance of economies. For two reasons in particular, this is disappointing. First, among all the 'laws' of economics, Engel's Law has probably received the most consistent endorsement and, as a result, changes in output structure and productivity growth would appear to be intimately related. Second, the advance of economics as a discipline is largely dependent upon the progressive relaxation of restrictive assumptions, yet the productivity growth literature has persisted in retaining the implicit presumption that output structures do not change.[2]

This chapter considers the impact of the changing relative importance of agricultural and processed food commodities in final demand on the productivity performance of the UK economy over the period 1968 to 1984. To this end, Vertically Integrated Sector (VIS) models are used to provide estimates of efficiency gains in the delivery of commodities to final demand, and of changes

1 Among the industries studied agriculture has received a disproportionate amount of attention. Recent substantive contributions include Ball (1985), for the USA, Rayner *et al.* (1986) and Thirtle and Bottomley (1992) for the UK. Estimates for other industries, e.g. food processing, have appeared less regularly in the literature and have often been produced as a by-product of studies covering large groups of industries, e.g. Armstrong (1974), Peterson (1979), McDonald (1991) and O'Mahony and Oulton (1990). Studies specific to food processing in the UK are Mordue and Marshall (1979) and Balasubramanyam and Nyugen (1991).

2 An important exception to this observation is Wolff (1985).

in economic structure. These estimates are used to analyse the impact of changes in the structure and scale of final demand on the performance of the economy. Particular emphasis is attached to the influences of changes in the structure of deliveries to final demand, i.e. what might be termed the Engel's Law effects.

A decomposition analysis of aggregate efficiency gain estimates allows the measurement of the efficiency implications of different types of structural changes. In the context of the UK's food system this is especially useful given the substantial efficiency gains for the agricultural and food processing sectors, and the appreciable changes in the proportions of final demands accounted for by food products. The results indicate that changes in economic structure have had substantial impacts upon the overall performance of the UK economy, and therefore that the issue of changing economic structure and relative performance is a potentially important subject that requires more analysis. The chapter is arranged as follows. The measurement of total factor productivity growth from VIS models is outlined in the following section. In the third section, a rule for consistent aggregation is reported and decomposition procedures are developed that allow for changes in the structure of deliveries to final demand. Empirical content is provided in the fourth section, using data for the UK. Finally, the concluding section draws together the insights of the analysis, and comments on constraints and opportunities afforded by them.

Multi-sector models and measured efficiency gains

Input-output model

The standard Leontief input-output model for a closed economy may be expressed by a system of three equations:

$$\mathbf{q} = [\mathbf{I} - \mathbf{A}]^{-1}\mathbf{f} \quad [6.1]$$

$$\mathbf{x} = \mathbf{Bq} = \mathbf{B}[\mathbf{I} - \mathbf{A}]^{-1}\mathbf{f} \quad [6.2]$$

$$\mathbf{p}' = \mathbf{p}'\mathbf{A} + \mathbf{wB} \quad [6.3]$$

where $\mathbf{q}$ = the n - vector of gross outputs;
$\mathbf{f}$ = the n - vector of final demands;
$\mathbf{x}$ = the m - vector of primary inputs;
$\mathbf{A}$ = the $(n * n)$ matrix of intermediate input coefficients;
$\mathbf{B}$ = the $(m * n)$ matrix of primary input coefficients;
$\mathbf{I}$ = the $(n * n)$ identity matrix;
$\mathbf{p}$ = the n - vector of output prices;
$\mathbf{w}$ = the m - vector of primary input prices.

The first equation states that the vector of gross outputs (**q**) is a function of the inter-industry technology (**A**) and the vector of final demands (**f**). The second equation states that primary input requirements (**x**) depend on gross outputs and the primary input technology (**B**), and thus are a function of technology and final demands. Note that the determinants of actual deliveries to final demand are exogenous to the system. The third equation states that prices are a function of technology and input prices (**p** and **w**).

The organisational structure of input-output models is by industries, which implies that the structure is defined by reference to the gross output vector, and as such is oriented towards the analysis of the supply side of an economy. This is consistent with the supply side emphasis of an overwhelming proportion of the productivity growth literature, and is compatible with the monitoring of market interventions that seek to influence the performance of specific industries. But, within an input-output model, changes in the composition of deliveries to final demand are important (if somewhat obscured) forces that influence the structure of gross outputs and hence the scale of industries. While the forces determining the structure of final demands are exogenous to the system, and therefore may legitimately be excluded when the focus of attention is on the relative efficiencies of different industries, they are of concern when attention is turned to the responsiveness of an economic system to consumer preferences.

Vertically Integrated Sector model

In the Vertically Integrated Sector (VIS) model, the structure is organised by (vertically integrated) sectors that deliver commodities to final demand: that is, constructed with reference to the vector of final demands. A VIS is a hypothetical construct that nets out intermediate inputs, expressing them in terms of the ultimate quantities of primary inputs that they represent. Thus, a VIS utilises primary inputs, drawn (potentially at least) from all industries in a system to produce commodities that are delivered direct to final consumers. For each industry there is a corresponding VIS, but only a part of the industry's primary inputs are allocated to its own VIS: each VIS is composed of primary inputs actually employed by other industries, but which contribute to the production of that VIS's (final) output. While the models are conceptually very different, they are related. Assuming, for simplicity, that primary inputs are homogeneous, then integrating [6.1] into [6.2] gives

$$\mathbf{x} = \mathbf{Bq} = \mathbf{B}[\mathbf{I} - \mathbf{A}]^{-1}\mathbf{f} = \mathbf{Rf} \qquad [6.4]$$

where **R** = the (m * n) matrix of vertically integrated primary input coefficients, and thus r_{jk} is the total (direct and indirect) requirement of the jth primary input per unit of the kth commodity delivered to final demand. Similarly, by integrating [6.1] into [6.3], final product prices are shown to be a function of primary input prices and the vertically integrated coefficients

$$\mathbf{p}' = \mathbf{w}'\mathbf{R} \qquad [6.5]$$

Thus the VIS model is a system of two equations, and, as [6.4] and [6.5] demonstrate, the step between the alternative models is a simple, linear transformation.

Efficiency gains and vertically integrated sectors

Total factor productivity growth, or efficiency gains, are defined as the rate of growth of real outputs that are unaccounted for by the rate of growth of real (conventional[3]) inputs. For the economy as a whole, real outputs are defined as value-added (or (domestically produced) deliveries to final demand) and real inputs as primary inputs. Similarly, for a VIS real output is defined as deliveries to final demand, and real inputs as all primary inputs used by the VIS, regardless of the industry that ultimately employs the primary inputs. Consequently, pairs of input-output tables, expressed in terms of constant prices, contain all the information necessary to estimate both economy-wide and VIS efficiency gains, provided that they are supplemented by data on primary input quantities.

The VIS measure of the rate of efficiency gains for an economy is thus

$$d \ln \gamma = -(\mathbf{p}'\mathbf{f})^{-1}[\mathbf{w}'.d\mathbf{R}]\mathbf{f} \qquad [6.6]$$

which is identical, by construction of the VISs, to the expression for the industry based model.[4] Efficiency gains are represented solely by the weighted reduction in the vertically integrated technical coefficients, but by definition these changes encompass changes in both primary and intermediate coefficients. Furthermore, note that the relationship is based upon final demands rather than gross outputs,

---

3 The distinction between conventional and unconventional inputs has been used in studies seeking explanations of the rate of efficiency gains, e.g. estimates of the rates of return to agricultural research and development. Examples of inputs defined as unconventional are education and Research & Development services.

4 Formally, the relationship between the rate of productivity growth from an industry based model, $d\ln t$, is

$$d\ln t = -(\mathbf{p}'\mathbf{f}\,)[\mathbf{p}'d\mathbf{A} + \mathbf{w}'d\mathbf{B}]\mathbf{q}$$

but

$$d\mathbf{R} = d\mathbf{B}[\mathbf{I} - \mathbf{A}]^{-1} + \mathbf{B}[\mathbf{I} - \mathbf{A}]^{-1}\, d\mathbf{A}\; [\mathbf{I} - \mathbf{A}]^{-1} = [d\mathbf{B} + \mathbf{R}d\mathbf{A}][\mathbf{I} - \mathbf{A}]^{-1}$$

and hence

$$d \ln g = -(\mathbf{p}'\mathbf{f})^{-1}[\mathbf{w}'.d\mathbf{R}]\mathbf{f} = -(\mathbf{p}'\mathbf{f})^{-1}[\mathbf{w}'d\mathbf{B} + \mathbf{w}'\mathbf{R}d\mathbf{A}][\mathbf{I} - \mathbf{A}]^{-1}\mathbf{f}] = -(\mathbf{p}'\mathbf{f}\,)[\mathbf{w}'d\mathbf{B} + \mathbf{p}'d\mathbf{A}]\mathbf{q} = d\ln \mathrm{t}$$

since intermediate inputs have been netted out by the process of vertical integration.

Similarly, measured productivity growth in the *k*th VIS is

$$d \ln \gamma_k = -\sum_{j=1}^{m} \frac{w_j . dr_{jk}}{p_k} \qquad [6.7]$$

whereby the change in the efficiency with which the *k*th commodity is supplied to final demand is represented by a weighted change in the vertically integrated coefficients.[5]

The transition between a closed and open economy VIS model is complex. Complications arise through the need to convert imported intermediate inputs, which are endogenously determined, into quantities of primary inputs, but imported intermediates are the product of non-domestic primary inputs. One solution is to represent imported intermediates in terms of the domestic primary inputs that would have been required to produce the (compensating) exports needed to purchase the imports, and then proceed to estimate efficiency gains as for a closed economy. This approach has been used to derive the VIS efficiency gains reported in this chapter. It is not described in detail, since the principal concern here is the aggregation of efficiency gains: this is not influenced by the procedure used to derive VIS efficiency gains themselves.[6]

Measured efficiency gains and aggregation

An important facet of measured efficiency gains is that of aggregative consistency. Specifically, it should be possible to aggregate estimates of efficiency gains by sectors/industries so as to (indirectly) derive estimates of economy-wide efficiency gains, and such estimates should equal those derived directly. This is an important topic that was relatively neglected in the productivity growth literature until recently.[7] However, in industry-based models this necessitates addressing a substantive problem: industry-level estimates have sought, by design, to distinguish between efficiency gains that originate within an industry from those gains that impact upon an industry (see Hulten, 1978, and Rayner *et al.*, 1986). Thus, estimates of efficiency gains by an industry should exclude any efficiency gains realised in the production of the intermediate inputs

5 The relationship between industry and VIS estimates of efficiency gains is developed in McDonald *et al.* (1992).

6 The procedure is reported in McDonald (1991), and derives from previous work by Schulter and Beeson (1979) and Rayner *et al.* (1986).

7 An important exception to this statement was work by Domar (1961).

that industry utilised, and hence, when aggregating industry estimates, it is necessary to reincorporate such efficiency gains.[8]

These difficulties do not arise in a VIS model. The process of creating VISs entails netting out inter-industry transactions, and therefore estimates of efficiency gains achieved by VISs represent the total change in the efficiency with which commodities are delivered to final demands. Thus, the problem of reincorporating efficiency gains transmitted through intermediate inputs does not arise in the VIS model, and aggregation of sectoral rates of efficiency gain can proceed by using the value shares of deliveries to final demand, i.e.

$$d \ln \gamma = \sum_{k=1}^{n} \left[\frac{p_k f_k}{\mathbf{p'f}} \right] d \ln \gamma_k \qquad [6.8]$$

A decomposition

The aggregation weights described in [6.8] are rooted in a continuous time model. Practical empirical work is constrained, however, to utilising discrete time data and the pragmatic solution to this problem, following a precedent set by Leontief (1953), has been to use weights calculated using the means of the relevant variables for the two time periods under immediate consideration. However, this solution achieves its objective by suppressing implications of changes in the final demand structures as an economy grows, since changes in an economy's output structure will cause changes in the weights themselves.

More specifically these changes reflect changes in the relative importance of different commodities in total deliveries to final demand. To disentangle the various effects it is useful to note that a vector of deliveries to final demand can be written as the product of a scalar which identifies the aggregate value of deliveries to final demand, and a vector which identifies the structure of the vector of final demands:

$$\mathbf{f} = (\mathbf{p'f})\mathbf{s} \qquad [6.9]$$

where $\mathbf{s} = \{s_i\}$ = an n - vector of value shares of deliveries to final demand:

$$s_i = \frac{f_i}{\sum_{i=1}^{n} f_i}$$

8 Two aggregation procedures have been proposed to solve this problem: the first was developed by reference to deliveries to final demands (Domar, 1961), while the second was developed by reference to values added (Jorgenson *et al.*, 1987). The procedure developed by Domar has been used by Peterson (1979), Wolff (1985) and McDonald (1991), while that developed in Jorgenson *et al.* has been used by Jorgenson *et al.* (1987) and O'Mahony and Oulton (1990).

Consequently, changes in the vector of final demands can be decomposed into changes in the total value of deliveries to final demand, that is, an aggregate growth component, and changes in the structure of deliveries to final demand, which may be termed an Engel's Law effect:

$$d\mathbf{f} = d[(\mathbf{p}'\mathbf{f})\mathbf{s}] = (\mathbf{p}'d\mathbf{f})\mathbf{s} + (\mathbf{p}'\mathbf{f})d\mathbf{s} \qquad [6.10]$$

The decomposition of changes in shares of deliveries to final demand can be readily developed from [6.10]. In particular, the deliveries to final demand by the kth VIS can be written as

$$f_k = (\mathbf{p}'\mathbf{f})\mathbf{s}_k \qquad [6.11]$$

where $\mathbf{s}_k$ is a vector of zeros except for the kth element which is the (value) share of the kth commodity in total deliveries to final demand. Then, the change in the quantity of the kth commodity delivered to final demand can be written as

$$df_k = (\mathbf{p}'d\mathbf{f})\mathbf{s}_k + (\mathbf{p}'\mathbf{f})d\mathbf{s}_k \qquad [6.12]$$

and thus the decomposition of the final demand share aggregation weights is[9]

$$d\left(\frac{f_k}{F}\right) = \frac{-(dF)f_k}{F^2} + \frac{F((\mathbf{p}'\mathbf{f})d\mathbf{s}_k)}{F^2} + \frac{F((\mathbf{p}'d\mathbf{f})d\mathbf{s}_k)}{F^2} \qquad [6.13]$$

where $F = (\mathbf{p}'\mathbf{f})$, i.e., the value of total deliveries to final demand.

Results

Data[10]

The data used in this study are two pairs of commodity by commodity input-output tables for 1968 and 1974 and for 1979 and 1984, expressed in constant 1970 and 1980 prices respectively. Commodity by commodity tables are available for both domestic and imported intermediate inter-industry transactions in each year. For the earlier period, 1968 and 1974, the tables are for a 29 industry model of the UK economy, whereas for the later period, 1979[11] and 1984, the tables are for a 28 industry model. The degree of aggregation was

9 Using the quotient rule, and noting, with constant prices, that the price vector can be treated as a numeraire.

10 Full details of the procedures used to compile the data series, and the data, are reported in McDonald (1991).

11 The 1979 tables were published in a form compatible with the 1980 SIC: see CSO (1983).

limited by the availability of capital stock data. The tables were derived from published (CSO) input-output tables for the UK, and it should be noted that two different Standard Industrial Classifications (SIC), SIC (1968) and SIC (1980), were used by the CSO. Thus, it is not possible to compare estimates meaningfully across the two pairs of tables, since the change in industrial classification scheme was substantial. Crucially for the purpose of this analysis, this included a change of name and scope of post-farmgate processing: the Food Processing sector was renamed Food, and there was a reallocation of Animal Slaughtering, and associated activities, from Distribution to Manufacturing, which (given the treatment of distribution services in the input-output tables[12]) affects the agricultural and the processing industries and the relevant VISs.

Three categories of primary inputs are identified: labour, capital and land. The technical coefficients for the primary inputs were derived from data on the 'physical' quantities of labour (in person hours), capital (in gross capital stocks) and land (in hectares), although land was only recorded as a separate input for Agriculture. Data for the rates of return to primary inputs were taken from the published input-output tables. However, since land is not identified as a recipient of payments in the input-output tables, it was necessary to adjust the reported profits for agriculture to allow for payments to land. This was achieved by using the MAFF land rent series.

Vertically integrated sectors and estimated efficiency gains

The impact of changes in the structure and aggregate level of deliveries to final demand upon the performance of the UK's economy in the production of final commodities can be seen from Table 6.1. For the earlier period, 1968-74, the changes in the aggregation weights between the first (base) year and the second (current) year made a substantial difference to the estimated rate of aggregate productivity growth, an increase of nearly a third of a per cent per annum. By way of contrast, for the later period, 1979-84, the impact was minimal, at less than 0.01 per cent per annum, and is well within the range of measurement error. Despite the minimal impact during the latter period, it would appear that changes in the structure of output have potentially important implications for the revealed productivity growth performance of economies. A feature of these estimates is that for the earlier period, changes in output structure were such as to result in a greater measured rate of efficiency gains in the current period structure than in the base period structure. This result was reversed for the later period.

---

12 To avoid all sales passing through the distribution industries, UK input-output tables have been compiled using the convention that distribution services are purchased by industries, rather than having industries sell their output to distributors.

Table 6.1. Output structure and the contributions of the agricultural and food processing/food sectors to the UK economy.

		1968-74		
Contributions	*Efficiency Gains*	*Base Weights*	*Current Weights*	*Per cent Change*
Agriculture	1.423	0.037	0.031	-15.1
Food Processing	0.501	0.037	0.032	-12.2
UK		0.152	0.469	208.5
		1979-84		
Contributions	*Efficiency Gains*	*Base Weights*	*Current Weights*	*Per cent Change*
Agriculture	1.954	0.017	0.026	55.3
Food	0.816	0.057	0.054	-5.5
UK		0.511	0.502	-1.6

This is borne out when attention is turned to the effects of changes in output structure on the impact of VISs and, in turn, upon the performance of the UK economy. Despite the improvements in the efficiency with which agricultural and processed foods were delivered to consumers during the period 1968-74, the change in output structure would have reduced their contributions to the economy's performance by 15 and 12 per cent respectively. But for the later period, although both food system sectors still showed substantially above average rates of efficiency gains, the effects of changing structure were such that it caused a substantial increase in the agricultural sector's contribution, up by nearly 0.01 per cent per annum, while the food sector's contribution showed a small potential decline.

The decomposition analysis developed earlier provides insights into these changes in contributions. Table 6.2 reports the decomposition of the differences between the productivity growth performance of the UK and the contributions of the agricultural and food processing/food VISs, using base and current period aggregation weights (see equation [6.13], above). The first feature that stands out from these decompositions is the change in the sign on the aggregate growth effects between the two periods, i.e. the final row of estimates. This simply reflects the fact that between 1968 and 1974 the economy grew, whereas between 1979 and 1984 the economy shrank. More importantly, however, it influences the interpretation that should be placed upon the changes in the structures of deliveries to final demand: specifically the expected sign on the effects of Engel's Law will be reversed.

Table 6.2. A decomposition of the agricultural and food processing/food sectors' contributions.

	1968-74 Agriculture	1968-74 Food Processing	1968-74 UK	1979-84 Agriculture	1979-84 Food	1979-84 UK
Difference	-0.0055	-0.0045	0.317	0.0093	-0.0032	-0.008
Decomposition						
$\frac{-(dF)f_k}{F^2}$	-0.0065	-0.0066	-0.027	0.0003	0.0009	0.008
$\frac{F((\mathbf{p}'\mathbf{f})ds_k)}{F^2}$	-0.0045	-0.0037	0.258	0.0095	-0.0032	-0.008
$\frac{F((\mathbf{p}'d\mathbf{f})ds_k)}{F^2}$	0.0055	0.0058	0.083	-0.0004	-0.0009	-0.008

This is largely borne out by the third row of estimates in Table 6.2. In the earlier period, the effect of changes in the structure of deliveries to final demand was to reduce the contributions of both the agricultural and food processing sectors to the performance of the economy. Although, strictly, the expectation would be that deliveries of all food commodities to final demand should diminish as a proportion of total deliveries, it is possible that when more than one sector delivers food commodities the relative importance of one sector can increase, with the increase being more than offset by changes in deliveries by the other sector(s). This is in fact what happened during the later period. Agriculture's share of deliveries to final demand increased substantially, from 0.9 per cent to 1.3 per cent, which represented approximately a 50% increase in deliveries. Correspondingly, food's share declined. The net effect was a slight increase in the proportion of deliveries to final demand accounted for by food system commodities, which is what Engel's Law would predict. It is important, however, to treat this result with considerable caution. The very large increase in the delivery of agricultural commodities direct to final demand is extremely difficult to rationalise. It may in part have been accounted for by changes in consumer preferences towards less processed foods, but a strong suspicion remains that the majority of the change is a result of increased sales of agricultural commodities into intervention stores. Since the achievement of the sign predicted by Engel's Law is marginal, virtually any increase in intervention purchases must undermine the apparent consistency.

The differences between the effects of changes in the structure of deliveries to final demand between the two time periods is of some interest. In the earlier period the change in the structure of deliveries to final demand is the dominant factor explaining the difference in the aggregate rate of efficiency gains, i.e. 0.26

of the 0.32 per cent per annum difference, while the impact upon the contributions of the food system of changes in final demand structure are less important, although clearly substantial. By way of contrast, in the later period changes in final demand structure are far less important, in relative terms, at the aggregate level than they were at the sectoral level. No readily apparent, theoretically substantiated explanation appears to exist for this change, although the negative rate of aggregate growth in the later period would be an obvious candidate. Whatever the ultimate explanation, it is arguable that these results are sufficient to indicate that changes in the structure of final demands may be an important factor in the determination of aggregate productivity growth, and therefore an important subject for further investigation.

Concluding comments

Changes in the final demand structure of the UK economy between 1968 and 1974 were consistent with an enhancement of the rate of aggregate efficiency gains, but, between 1979 and 1984, were such as to reduce the rate of aggregate efficiency gains. This restructuring of economic systems in response to technological changes (broadly defined) and consumer preferences, appears to have been a neglected topic within the productivity growth literature,[13] even though it is clearly a fundamental characteristic of growing economies.

The process of restructuring entails a constant readjusting of the relative importance of directly and indirectly employed primary inputs in the production of final demands, while simultaneously altering the relative importance of different commodities in total deliveries to final demand. Furthermore, changing the patterns of interdependence provides a major incentive for the use of multi-sector models in the analysis of economic systems. Such models serve to highlight the importance of the distinction between how efficiently sectors utilise inputs: that is, the focus of attention in sectoral studies of productivity growth, and how a sector contributes to the efficiency with which the system as a whole performs.

Evidence of these effects has been provided in this paper with respect to an agricultural sector and a monolithic Food Processing/Food sector. The analysis has been constrained, however, by both the high degree of aggregation and the lack of an extended series of input-output tables on a common system of classification. The publication of 1990 input-output tables for the UK will go some way towards addressing the second constraint.

The first constraint is, however, potentially more interesting. Greater disaggregation of the Food Processing/Food sector offers advantages since it would allow the development of the distinction between food processing activities that are principally concerned with bulk processing and producing

---

13 Indeed, it is arguable that since the productivity growth literature has been based largely upon the concept of an aggregate production function that such have been excluded by assumption.

largely undifferentiated products, and those activities principally producing low volume, high value food commodities. The current Standard Industrial Classification system, and the degree of disaggregation in its adaptation by published input-output tables, are to some extent consistent with this broad classification. Thus, should sufficient capital stock data become available, analysis in more detail will be possible, with the consequence of potentially greater insights into the effects of changes in output structure upon overall economic performance.

References

Armstrong, A.G. (1974). *Structural change in the British economy 1948-1968, Volume 12 of a programme for growth*, Cambridge and London: Department of Applied Economics (University of Cambridge) and Chapman and Hall.

Balasubramanyam, V.N. and Nyugen, D.T. (1991). Structure and performance of the UK food and drink sector, *Journal of Agricultural Economics*, 42, 56-65.

Ball, V.E. (1985). Output, input and productivity measurement, *American Journal of Agricultural Economics*, 67, 475-486.

Central Statistical Office (CSO) (1983). *Input-output tables for the United Kingdom 1979*, London: HMSO.

Domar, E. (1961). On the measurement of technological change, *Economic Journal*, 71, 709-729.

Hulten, C.R. (1978). Growth accounting with intermediate inputs, *Review of Economics and Statistics*, 45, 511-518.

Jorgenson, D.W., Gollop, F.M. and Fraumeni, B.M. (1987). *Productivity and US economic growth*, Cambridge: Harvard University Press.

Leontief, W.W. (1953). *Studies in the structure of the American economy*, New York: Oxford University Press.

McDonald, J.R.S. (1991). Productivity growth and structural change in the UK economy and food system: 1954-84, unpublished PhD Thesis, University of Nottingham.

McDonald, J.R.S., Rayner, A.J. and Bates, J.M. (1992). Productivity growth and the UK food system, *Journal of Agricultural Economics*, 43, 191-204.

Mordue, R.E. and Marshall, J.P. (1979). Changes in total factor productivity in UK food and drink manufacturing, *Journal of Agricultural Economics*, 30, 159-166.

O'Mahony, M. and Oulton, N. (1990). *Growth of multi-factor productivity in British industry: 1954-86*, NIESR Discussion Paper, 182.

Peterson, W. (1979). Total factor productivity growth in the UK: a disaggregated analysis. In: K. Patterson and M. Schott (eds.) *The measurement of capital – theory and practice*, London: Macmillan, 212-225.

Rayner, A.J., Whittaker, J.M. and Ingersent, K.A. (1986). Productivity growth in agriculture revisited: a measurement framework and some empirical results, *Journal of Agricultural Economics*, 37, 127-150.

Schulter, G. and Beeson, P. (1979). Components of labour productivity growth in the food system, 1958-67, *Review of Economics and Statistics*, 63, 378-384.

Solow, R.M., (1957). Technical change and the aggregate production function, *Review of Economics and Statistics*, 39, 312-320.

Thirtle, C. and Bottomley, P. (1992). Total factor productivity in UK agriculture, 1967-90, *Journal of Agricultural Economics*, 43, 381-400.

Wolff, E.N. (1985). Industrial composition, interindustry effects and the US productivity slowdown, *Review of Economics and Statistics*, 52, 268-277.

Appendix

For information, the constant price values of deliveries to final demands and the shares of total deliveries to final demand for each commodity group are reported as Tables 6.A1 and 6.A2.

Table 6.A1. Deliveries to final demand and shares of total deliveries to final demand: 1968 and 1974.

	Final Demand		*Shares of Final Demand*	
Industry	*1968*	*1974*	*1968*	*1974*
Agriculture	1,075	1,112	0.026	0.022
Forestry & Fishing	98	68	0.002	0.001
Coal Mining	215	103	0.005	0.002
Other Mining & Quarrying	42	67	0.001	0.001
Food Processing	3,088	3,304	0.074	0.065
Drink & Tobacco	1,038	1,598	0.025	0.031
Coal & Petroleum	432	1,057	0.010	0.021
Chemicals etc.	1,268	2,160	0.030	0.042
Iron & Steel	326	347	0.008	0.007
Non-ferrous Metals	257	363	0.006	0.007
Mechanical Engineering	2,572	2,972	0.061	0.058
Instrument Engineering	365	572	0.009	0.011
Electrical Engineering	1,629	2,337	0.039	0.046
Shipbuilding	442	490	0.011	0.010
Vehicles	2,634	2,820	0.063	0.055
Metal Goods nes*	587	703	0.014	0.014
Textiles	1,178	1,285	0.028	0.025
Leather & Clothing	1,041	1,307	0.025	0.026
Building Materials	217	321	0.005	0.006
Timber etc.	448	512	0.011	0.010
Paper Products	737	780	0.018	0.015
Other Manufacturing	487	569	0.012	0.011
Construction	5,346	5,571	0.128	0.109
Gas	245	1,007	0.006	0.020
Electricity	750	882	0.018	0.017
Water	96	267	0.002	0.005
Transport & Communications	3,191	4,088	0.076	0.080
Distribution	5,748	6,503	0.137	0.128
Services nes*	6,297	7,830	0.150	0.154
Total	41,850	50,993		

* nes – not elsewhere specified.

Table 6.A2. Deliveries to final demand and shares of total deliveries to final demand: 1979 and 1984.

	Final Demand		*Shares of Final Demand*	
Industry	*1968*	*1974*	*1968*	*1974*
Agriculture	1,760	2,690	0.009	0.013
Forestry & Fishing	276	268	0.001	0.001
Coal Extraction etc.	609	-100	0.003	0.000
Extrn. of Min. Oil and Natural Gas	4,893	8,320	0.024	0.041
Mineral Oil Processing	5,623	4,130	0.028	0.021
Electricity & Nuclear Fuel	4,143	4,647	0.020	0.023
Gas	2,253	2,535	0.011	0.013
Water Supply	468	474	0.002	0.002
Metals	3,047	2,827	0.015	0.014
Other Mins. & Mineral Products	1,486	1,155	0.007	0.006
Chemicals & Man Made Fibres	8,103	9,268	0.040	0.046
Metal Goods nes*	2,231	1,809	0.011	0.009
Mechanical Engineering	13,557	8,738	0.066	0.043
Electrical & Instr. Engineering	10,063	11,400	0.049	0.057
Motor Vehicles & Parts	7,904	6,153	0.039	0.031
Other Vehicles	5,193	5,285	0.025	0.026
Food	14,287	13,281	0.070	0.066
Drink & Tobacco	4,910	3,695	0.024	0.018
Textiles	2,757	2,322	0.013	0.012
Clothing, Footwear & Leather	4,281	3,580	0.021	0.018
Timber & Wooden Furniture	2,495	2,291	0.012	0.011
Paper & Printing	3,132	2,879	0.015	0.014
Rubber & Plastic	1,834	1,482	0.009	0.007
Other Manufacturing	1,747	929	0.009	0.005
Construction	23,477	22,325	0.115	0.111
Distribution	41,559	44,541	0.203	0.222
Transport & Communication	12,843	11,334	0.063	0.056
Services	19,333	22,733	0.095	0.113
Total	*204,262*	*200,990*		

* nes – not elsewhere specified.

Chapter seven: **Problems with Leontief Technology and Agriculture: the Case of Producing Commodities with More than One Input**

Alastair Bailey

Department of Agricultural Economics, Wye College, The University of London

Introduction

This chapter is an attempt to draw researchers' attention to some implicit theoretical restrictions assumed by Leontief's input-output approach. In an attempt to keep discussion concise, attention is directed specifically toward the implications of inference drawn from the estimation of input-output coefficients from whole farm data using Errington's (1989) extension to Tyler's (1966) technique.

The chapter is set out as follows. Firstly, the theoretical implications of assuming that multiple input production processes are Leontief-characteristic and the effects of multi-output production are explored. Secondly, the practical implications of finding these implicit assumptions to be invalid are discussed in the context of prediction of economic inter-sectoral linkages: here, specific reference to agriculture and the rural economy is made. Thirdly, a more general specification of the production technology than Leontief's original specification is suggested. Finally, the discussion in the three major sections is drawn together in order to make some conclusions.

To provide some context for the major foci of this chapter, consider Errington's (1989) suggestion that the individual input-output coefficients, a_{ij}, could be estimated from Farm Business Survey data. The accounts do incorporate sufficient spatial information to form rough and ready regional estimates of these technical coefficients. The estimation technique utilises stochastic properties of the general linear statistical model to provide sample estimates of the a_{ij}s, for each input i and output j of the agricultural sector. Its specification assumes that each input i (in quantity or value terms) is endogenously determined by a set of n independent linear regression equations where output j (again in quantity or value) is assumed to be exogenous of the data generation process. Midmore (1991b) has discussed the various econometric difficulties involved in this approach.

Theoretical restrictions of Leontief production technology

The major issue concerning the application of the input-output approach to agriculture rests on the highly restrictive technological structure imposed upon the production within each sector.

The coefficients a_{ij} of the technical matrix **A** describe the ratio of the amount of input i required to produce one unit of output j. The estimation of elements of **A** from cross-section data (see Tyler, 1969; Errington, 1989) has two major implications for the subsequent analysis. Firstly, the coefficient matrix depicts the technology point for the production at the mode output level, and secondly, the coefficients a_{ij} represent the mean technology t from the set T, where T is the set of all available and technologically feasible combinations in input-output, input-input space. T thus resides in $Ij + II$ space, where $I = \Sigma i$. For the case of multi-output production T also includes relationships in output-output space. Here T is in the real space $IJ + II + JJ$, where $J = \Sigma j$.

The first point above has importance from the point of view of the orientation of study. Taking agriculture as a relevant case, the coefficients matrix describes the input-output relationships, biased toward a modal farm size. From an *a priori* knowledge of the size distribution of farms, the expectation would be that this mode farm size would be within the lower end of the skewed size distribution of holdings. From the point of view of analysis of inter-sectoral employment implications, this result is clearly advantageous. However, if the objective is to describe differing impacts of change upon different sizes of farms it is seriously flawed. Equally, if the focus is on impacts of policy changes on the output of a region, this point is also of importance. The size distribution of farms in the UK, for example, suggests that approximately 25 per cent of all farms produce three-quarters of all output. The Leontief production function assumes constant returns to scale, without testing for it: yet econometric studies (for example Dawson and Hubbard, 1987, or Mukhtar and Dawson, 1989) have reported the existence of 'L' shaped, 'U' and skewed 'U' shaped average cost curves for UK agriculture.

The second point above refers directly to the theory of convex sets. The production possibility set T is said to be closed, that is, bounded, and convex. The choice of notation here is arbitrary, since the set T is commonly associated with the profit function: however, similar arguments can be used to discuss the convexity of $V(y)$, the input requirement set for single output production functions and cost function (see Chambers, 1988, for greater detail). Boundedness implies that the set is closed, or that any point outside the set is infeasible. Convexity implies that if two points t_1, t_2 are elements of the set T for some output level (or profit level), then any weighted average of these two points can also produce that output level. If these two points lie on the boundary of *T*, then it is clear that concavity of *T* is excluded. The set *T* resides in the *n* space scribed by the *I* inputs and *J* outputs ($i = 1...I, j = 1...J$) of that sector.

Assume, for simplicity, that $I = 2$ and $J = 1$: this depicts a standard neoclassical unit isoquant. Recall that all points which lie upon the isoquant (in the case of *T*, its boundary) represent technically efficient points, whilst those which lie above the frontier, or within the body of the set *T*, are inefficient. Consider now how the elements of the set *T* relate to the nature of farm-level cross-section data. Each farm commits some combination of inputs, both primary and secondary, and uses up some proportion of its stock of capital goods to produce outputs. However, each farm in any population may be at differing points in the life of their capital stock or technology. Thus a continuum of technology points should emerge and, in addition to random shocks, should give rise to the mapping of *t* within the set *T*. For this reason estimates derived from cross-section data should be viewed as an underestimation to some long-run equilibrium state technology, made up of points within the boundary of *T*.

Returning attention to a consideration of the technology matrix **A** from Leontief's specification of the input-output model, considerable clarity is required concerning consequent constraints imposed on the estimated production process of each sector. The production structure assumed by Leontief and the majority who followed him is that of 'fixed proportions of factors of production'. Here factors of production are assumed to be combined with each other in fixed proportions to produce any given output. Should desired output rise, then the function which produces it dictates that each factor of production should rise in proportion with output.

Leontief (1951) dismissed the importance of substitution between two or more inputs. This assumption rests upon the observation that manufacturing industries employ relatively inflexible capital technologies: '... such expenditures are determined by relatively inflexible engineering considerations or by equally inflexible customs and institutional arrangements ... ' (p. 22). Since Leontief was concerned with modelling large-scale economic systems (predominantly at national or international level), this broad brush assumption was, in some ways, reasonable. In addition, Leontief used national income accounts data from which to calculate his technical coefficients. This information source includes only industry/sector input and output levels in either quantity or value units for the period. Hence these coefficients, while providing best estimates of current production practice, can do no more than indicate the current relationship between inputs and outputs for the prevailing price ratios and current state of technology. Even if the majority of these technical relationships are assumed to exhibit elasticities of substitution which, *a priori*, may be considered close to or equal to zero, it does seem to be highly detrimental to impose such a restriction without testing for its validity.

The restriction of zero input substitution is also imposed by the Errington technique for estimating farm-level input-output coefficients. Here, imposition occurs at the estimation stage: independent estimation of each input demand function, in turn, constrains all elasticities of input substitution to be equal to zero.

Fixed-factor proportions and agricultural production

Midmore (1991a, 1993) has expressed scepticism about the validity of this assumption when the economic (sub)system under observation includes the agricultural sector. To address and explore these doubts further, initially the characteristics which distinguish agriculture from other sectors of the economy need to be discussed. Subsequently, the effects of these distinguishing characteristics on the production processes within agriculture are considered.

The role of land as a primary resource

Agriculture is a primary production sector. Finite primary resources are combined with labour and the products of other industries, including agriculture itself, to produce biologically based commodities as output to other sectors.

The key point of interest here is that the primary resource, land, is finite in quantity. Land improvement is possible, but in general a gain in quality may only be achieved by committing a greater quantity of other resources to the available land area; output can be expanded by increasing the commitment of, say, fertiliser or investment in drainage to a given land area. If this scenario is considered from the point of view of holding output constant, then 'land improving inputs' are substituted for land. The relationship described here is one of complementarity. The marginal rate of technical substitution (MRTS) lies between zero and one. The Leontief fixed-factor technology restricts the MRTS to be equal to zero, and is thus nested. However, if the MRTS *does* equal zero, and if the fertiliser or drainage input is zero, then output should be zero, as all inputs are perfect complements and are strictly necessary for positive outputs.

Multi-output production

A further important characteristic which distinguishes agriculture from other productive sectors of the economy is that of the production of multiple outputs. For completeness, agricultural production can be defined as multi-output and multi-input, since in general more than two inputs are combined in such a way as to produce an array of outputs.

Economies of scale and multi-output production

The production of more than one output is not incidental to the structure of production. Resorting to neoclassical micro-economic theory and the assumption of profit maximisation, the occurrence of this phenomenon can be explained. If production is 'non-joint', then the only explanation for the production of multiple outputs is that single output production of one or more commodity exhibits diseconomies of scale. For this to hold, the restrictions imposing Constant Unit Returns to Scale and the MRTS to equal zero must be assumed to be invalid, since as output rises the cost function is increasing for at least some output range.

The case of 'jointness' in production

The above mentioned concept of 'non-joint' production is the opposite of 'jointness' in production. For jointness to hold there must be some cost advantage to the production of two or more commodities within the same firm.

Shumway *et al.* (1984) suggest that, if jointness in the production of multiple outputs can be assumed, then the specification of a single equation multi-output production function is consistent with neoclassical theory. As a special case, joint production includes those products which are biologically linked, such as grain and straw, or milk and dairy calves. However, it may be possible for other, less formal, cost saving (and therefore output augmenting) relationships to satisfy these sufficiency criteria. Here, the possibility that fixed but allocatable inputs can cause jointness in production is discussed. Leathers (1991) used a multi-product cost function approach developed by Baumol *et al.* (1981) to discuss the causes of jointness in US agriculture. In conclusion, they suggested that jointness in production can be caused by the presence of allocatable fixed inputs only in the short run. However, this cause of jointness does not imply that joint production will be optimal, as the existence of economies of scale for single output production could outweigh its effects.

Reverting now to the problem being considered in this section, and recalling that the estimates, a_{ij}, were conceived of as long-run approximations to the input-output coefficients, the following inference can be made. Since, in the long run, the neoclassical theory of the firm assumes that all inputs are variable with the exception of land (which is only variable in the very long run, and for the firm only), allocatable fixed inputs are implicitly ruled out as a cause of jointness. In addition, the assumption of fixed proportions of factors of production implicity assumes that returns to scale will equal unity for each product. Both these assumptions are then inconsistent with empirical and *a priori* observation. The fact that the majority of farms are clearly multi-product in the long run contradicts the Leontief technology assumptions.

The implications of assuming 'fixed factor/product proportions' for analysis both within and between sectors

From the discussion so far it is clear that the simplifying assumptions imposed by the Leontief input-output model and (in turn) Errington's ad hoc estimation procedures, place severe restrictions upon inference. These restrictions become more important as the level of detail increases, especially where multi-product agriculture is the subject of study.

The model, as it stands, restricts returns to scale to be constant at unity. Thus, within a geographical region of finite size, any trend toward increasing farm size (made possible through the amalgamation of existing holdings) would have no effect, either on the level of final output or on production costs associated with it. If this was so, then the optimal farm size for that region would be the area of that region itself. This latter point would be brought about by some income effect as no economic optimum size is defined.

The technical coefficients matrix **A** should thus be regarded as simply a very local approximation to the average technology for the mode farm size. The properties of constant returns to scale at unity, and zero elasticities of factor and product substitution have been imposed by prior assumption and no test for this property has been made. Therefore, the measurement of the effects of a change in final demand upon this or any upstream sectors should only be attempted for very small relative movements in demand or supply.

If the objective is to assess the impact of a relative scarcity of factor i, then by imposing the elasticity of substitution to equal zero, the result that output must fall by the same proportion is also imposed. The implication of this restriction is to negate the possibilities of input substitution, since all inputs are regarded as strictly necessary in their present observed proportions. Reducing the input of one factor by some proportion implies that output also falls by that proportion, and all other inputs are in slack by that same proportion. The neoclassical law of diminishing returns is violated.

In reality, it is equally, if not more likely that output would fall less rather than in equal proportion to the fall in supply of factor i. If this were the case, the assumption of fixed proportions of factor inputs will overestimate the effect on final output and on the demand for other factors of production.

There is also the implicit assumption that the farm sector is in equilibrium at the time of coefficient estimation. Consequently, it must also be assumed that the observed factor proportions are also technically optimal. As a result of the constraint that elasticities of factor substitution equal zero, changes in factor price ratios result in unchanged factor proportions, and no economic optimum factor demands are defined. Thus, observed input-output relationships are snapshots, specific to the state of technology, input price ratios, and the structure of firms within the sector.

The basic input-output model can be modified to accommodate some level of dynamic structural adjustment to investment in technical progress: see Leontief

(1966), and Midmore (1991a) for a discussion of the possibilities of using Mathur's (1977) vintage capital technique. Nevertheless, the problem of adjustment in response to changing input price ratios remains.

The problem of multi-output production is also of importance in assessing any response to change in relative prices. Given that farmers can change the combination of outputs they produce, then it should be expected that output price ratio changes should induce changes in the composition of total output. In addition to simple output price ratio changes, consider the case of changing the price, *ceteris paribus*, of some output-specific input. The increase in the price of such a factor should, according to neoclassical theory, precede two reactions. Firstly, input-input substitution should occur, reducing but not fully cancelling out the effect upon cost (given that the cost function for this output is well behaved, in the sense that it is non-increasing in input prices). Secondly, the relative profit from producing output i over the next most profitable alternative output j will decrease. Thus, in this case, not only does the coefficient overestimate the effect of change upon the output i and its associated costs, but the method also ignores any effect upon the production of output j. Thus, though not explicit, the input-output model will presuppose input non-jointness.

Suggestions for a more general specification of the production structure: the input-output model extended

Techniques exist within the mainstream of production economics and econometrics which can provide a much more general specification of the underlying production technology. Developments in the specification of so-called 'flexible functional forms' (see Diewert, 1971), which impose fewer prior restrictions on production relationship measurement, are now in common usage. It is beyond the scope of this chapter to discuss relative merits of the various forms which can claim to be second order Taylor series approximations. However, such functional forms do not place prior restrictions upon the elasticity of factor substitution, and, despite the fact that some prior restrictions are placed on the production relationship itself, these 'standard regularity restrictions' are rarely contrary to theory.

Concurrently, production economics saw the development and application of the dual approach to analysis. Duality theory offers a simple method of measuring the system of derived demands and the output supplies directly from a 'dual objective function' expressed in prices and quantities. It provides the opportunity to capture all the economic and technical information from the cost ($c(w,Y)$) or profit function ($\Pi(p,w)$).

From the cost function, estimates of the derived demands for factor inputs and their elasticities, with respect to own and cross prices, can be obtained. An extension to the multi-output cost function also allows the solution for optimal output supply bundles to changes in response to both output and input prices. The general suggestions which follow are, for simplicity, restricted to the estimation of cost functions. This restriction confines the analysis of own-price and cross-price elasticities to that of the Hicksian compensated variety. This type of price elasticity assumes that the aggregate level of the output bundle remains constant and costs are minimised accordingly.

Since the technical coefficients, a_{ij}, in the input-output model describe average factor cost, and since they are used to predict total demand for factor i and total output, that convention will be followed here. For the cost function, cost minimisation for the production of the exogenous aggregate output y equates to profit maximisation. This is achieved by equating marginal factor cost of producing Y to the price of that factor, for each factor.

The generalised Leontief single output cost function

Using Diewert's 'Generalised Leontief Cost Function', assuming that output Y and input prices w_i are exogenous and that input quantities x_i and total costs C are endogenously determined we can specify, for $i = 1,....,n; j = 1,....,m$:

$$C = Y.[\Sigma_i \Sigma_j b_{ij} (w_i w_j)^{½}] \qquad [7.1]$$

Constant returns to scale of the dual production function are imposed, as is symmetry ($\beta_{ij} = \beta_{ji}$). Optimal cost minimising factor demands can be obtained by applying Shephard's Lemma. Thus differentiating the generalised Leontief cost function in [7.1], with respect to w_i, yields, for $i = 1,....,n$:

$$\frac{\delta C}{\delta w_i} = x_i = Y.\Sigma_i b_{ij} (w_j / w_i)^{½} \qquad [7.2]$$

Dividing [7.2] through by Y yields the average factor cost function. This is synonymous with Errington's derived demands for inputs, yet this approach allows substitution, and so an optimum is defined for exogenous input prices. For this functional form, the average factor demands may be estimated as:

$$\alpha_i = \frac{x_i}{Y} = \Sigma_i b_{ij} (w_j / w_i)^{½} \qquad [7.3]$$

Estimation is simultaneous over n input demand functions. Each demand function includes n+1 parameters, one own-factor demand, β_{ij}, and n cross-price elasticities, β_{ij}. With the imposition of symmetry, the number of free parameters falls from $n(n+1)$ to $n((n/2)+1)$.

The estimated parameters of this approach are compatible with the input-output model of the sector. Those derived from [7.3] above form the diagonal matrix **B** of cross-price elasticities β_{ij} and own-price elasticities β_{ii}. The **A** matrix is thus the product of **B** and the price matrix **W** whose elements are w_i / w_j, and a vector of price changes.

The translog specification: the single output case

The translog specification of the above problem is somewhat less straightforward. However, the level of generality it does afford is very attractive. As with all so-called flexible functional forms, no *a priori* restrictions are placed upon input substitution. In addition, though, constant returns to scale need not be imposed and the degree of RTS is testable. The translog cost function is

$$\ln C = \ln\alpha_0 + \Sigma_i\, \alpha_i \ln P_i + \tfrac{1}{2} . \Sigma_i\Sigma_j\, \gamma_{ij} \ln P_i \ln P_j + \alpha_y \ln Y + \tfrac{1}{2}\, \gamma_{yy}(\ln Y)^2 + \Sigma_i\, \gamma_{iy} \ln P_i \ln Y \qquad [7.4]$$

To gain the optimal input demand equations from this functional form it is necessary to first logarithmically differentiate with respect to input prices, and then to apply Shephard's Lemma, as follows:

$$S_i = \frac{\delta \ln C}{\delta \ln w_i} = \frac{w_i}{C} \cdot \frac{\delta C}{\delta w_i}$$

$$S_i = \alpha_i + \Sigma_j\, \gamma_{ij} \ln w_j + \gamma_{iy} \ln Y$$

where;

$$C = \Sigma_i\, w_i\, x_i$$

$$S_i = \frac{w_i\, x_i}{C}$$

and therefore,

$$\Sigma_i\, S_i = 1 \qquad [7.5]$$

The share equations, S_i, above contain own effects, cross-price substitution effects and output effects. The cross equation constraints imposed by symmetry through the cross-price elasticities provide overidentification. Thus the above specification is estimated as the total cost function in [7.4], along with n-1 share equations from [7.5]; the coefficients of the omitted share i can then be inferred from symmetry and adding up.

This system yields share equations which look remarkably similar to the estimable factor demands of the generalised Leontief cost function. However, these cost share equations describe the optimal factor demands as the share of factor i of total cost. They are not analogous to the input-output relationship required, since they are not expressed over Y.

This problem can be solved by simply dividing the total cost function in [7.5] by output Y. This yields the 'unit' or 'average cost function' which is a function of the same variables as total cost and, as such, does not alter the specification. Thus, the following is estimated:

$$\ln LAC = \ln\alpha_0 + \Sigma_i\, \alpha_i \ln w_i + ½ \,.\, \Sigma_i \Sigma_j\, \gamma_{ij} \ln w_i \ln w_j + \alpha_y \ln Y + ½\, \gamma_{yy} (\ln Y)^2 + \Sigma_i\, \gamma_{iy} \ln w_i \ln Y$$

$$S_i = \frac{\delta \ln LAC}{\delta \ln w_i} = \frac{w_i}{LAC} \cdot \frac{\delta LAC}{\delta w_i} = \frac{w_i x_i}{LAC}$$

$$= \alpha_i + \Sigma_i\, \gamma_{ij} \ln w_j + \gamma_{iy} \ln Y \qquad [7.6]$$

The 'adding up' condition, that the sum of the S_is should equal one, implies that each S_i denotes the ratio of the derived demand for factor i to the average cost of producing one unit of output. S_i therefore describes how the minimum average cost should be apportioned between each factor to maximise profits for exogenous output level Y.

The profit maximising, cost minimising, derived demand for factor i can thus be calculated by simply multiplying average cost, or unit cost (the function value ln LAC) by S_i to obtain optimal estimates of a_{ij}.

Should separate estimates of the own-price elasticities and cross-price elasticities be required, then decomposition of S_i is necessary. Here decomposition follows Kako (1978) and Binswanger (1978) who used this form of decomposition to discover the bias of technical change in agriculture for Japan and the US respectively. Equations [7.7] and [7.8], below, provide details of how this may be achieved.

$$\frac{\delta S_i}{\delta w_i} = \frac{\delta\,(\alpha_i + \Sigma_i \gamma_{ij} \ln w_i + \gamma_{iy} \ln Y\,)}{\delta w_i}$$

$$\frac{\delta S_i}{\delta Y} = \frac{\delta\,(\alpha_i + \Sigma_i \gamma_{ij} \ln w_i + \gamma_{iy} \ln Y\,)}{\delta Y} \qquad [7.7]$$

If the demand for factor $i = LAC(w,Y) \,.\, S_i$, then:

$$\frac{\delta x_i}{\delta w_i} = LAC(w,y) \cdot \frac{\delta\,(\alpha_i + \sum_i \gamma_{ij} \ln w_i + \gamma_{iy} \ln Y)}{\delta w_i}$$

$$\frac{\delta x_i}{\delta Y} = LAC(w,y) \cdot \frac{\delta\,(\alpha_i + \sum_i \gamma_{ij} \ln w_i + \gamma_{iy} \ln Y)}{\delta Y} \qquad [7.8]$$

For the 'non-homothetic' translog average cost function presented above, the scale term γ_{ij} ln w_i can also be decomposed to allude to the effect of RTS in the neighbourhood of Y.

The appropriate stochastic specification of the cost functions in [7.1], [7.4] and [7.6] is that the additive random error term μ be distributed normally with zero mean and variance σ^2. This standard specification has the advantage that the 'average' functional relationship is disclosed. The estimated function is thus not that on the boundary of *T*, but lies within *T* by σ^2. This allows us to make assumptions about the sample population and to integrate these estimates into the input-output framework.

Thus, following the procedure adopted for the generalised Leontief production function, separate own-effect vectors and cross-price matrices may be constructed. In addition though, the translog specification of the cost function (without homotheticity and homogeneity of constant degree in output imposed) allows a matrix of output effects to be constructed.

This approach could then be extended to estimate optimal output supply systems for each agricultural output, using the multi-output profit function with n-1 input share and m-1 revenue share equations and cross effects. The effect of exogenous changes in p_j and w_j can be traced through both input substitution and output substitution. In using the profit function, the elasticities of factor substitution are expressed as Marshallian and the optimal level of output is considered to be endogenously determined.

Integrating this sectoral model with similar models of other sectors would then provide estimates of inter-sectoral linkages which conform to theory, thus giving some idea of the level of overestimation the standard Leontief model imposes.

Conclusions

In this chapter, the restrictions implied by the application of Leontief input-output techniques have been criticised on theoretical grounds, with special reference to agriculture. The imposition of a production technology which can only exhibit constant returns to scale (CRS) at unity and elasticity of output and input substitution equal to zero does appear to be tenuous, in farming perhaps more than in other industries. The assumption of unit CRS gives rise to severe limitations for the analysis of sectoral structural change upon sector supply and demand. Constraining the estimated average input costs to conform to an elasticity of input substitution of zero implies that changes in differential prices should consistently overestimate any change in production costs or their effect on the level of sector output.

Two suggestions have been made to generalise the input-output approach of inter-sectoral modelling, relaxing firstly the elasticity of input substitution with $0 \leq CRS \geq 1$, and secondly by allowing returns to scale themselves to vary.

The fact that agriculture is a multi-product sector throws doubt upon the validity of Leontief's production technology and consequently poses difficulties for analysis. Assuming that different outputs are produced by subtly different, if interlinked, production functions, then changes in output price ratios do induce change in the bundle of goods produced. Such changes do then have implications for the farm's input requirements. In order to address this criticism it is suggested that a multi-output cost or profit function approach be applied, extending the proposed approach still further.

References

Baumol, W., Panzar, J. and Willig, R. (1981). *Contestable markets and the theory of market structure*, New York: Harcourt, Brace, and Jovanovich.

Binswanger, H.P. (1978). Measured biases of technical change: the United States. In: Binswanger, H.P. and Ruttan, X. (eds.) *Induced innovation: technological institutions and development*, Baltimore: John Hopkins University Press, 215-242.

Chambers, R.G. (1988). *Applied production analysis; a dual approach*, Cambridge: Cambridge University Press.

Dawson, P.J. and Hubbard, L.J. (1987). Management and size economies in the England and Wales dairy sector, *Journal of Agricultural Economics*, 38, 27-37.

Diewert, W.E. (1971). An application of Shepherd duality theorem: a generalized linear production function, *Journal of Political Economy*, 79, 482-507.

Errington, A.J. (1989). Estimating enterprise input-output coefficients from regional farm data, *Journal of Agricultural Economics*, 40, 52-56.

Kako, T. (1978). Decomposition analysis and derived demand for factor inputs: the case of rice production in Japan, *American Journal of Agricultural Economics*, 60, 628-635.

Leathers, H.D. (1991). Allocatable fixed inputs as a cause of joint production: a cost function approach. *American Journal of Agricultural Economics*, 73, 1083-1090.

Leontief, W.W. (1951). *Studies in the structure of the American economy 1919-1939*, Cambridge: Cambridge University Press.

Leontief, W.W. (1966). Input-output analysis. In: Leontief, W.W. (ed.), *Input-output economics*, New York: Oxford University Press, 134-155.

Mathur, P.N. (1977). A study of sectoral prices and their movements in an input-output framework. In: Leontief, W.W. (ed.), *Structure, systems and economic policy*, Cambridge: Cambridge University Press, 29-47.

Midmore, P. (1991a). Input-output and agriculture: a review. In: Midmore, P. (ed.) *Input-output models in the agricultural sector*, Aldershot: Gower, 1-20.

Midmore, P. (1991b). *Problems in estimating agricultural input-output coefficients by regression methods*, Aberystwyth: Department of Economics and Agricultural Economics, Aberystwyth Rural Economy Research Papers, No 91-02 (University of Wales).

Midmore, P. (1993). Input-output forecasting of regional agricultural policy impacts, *Journal of Agricultural Economics*, 44, 284-300.

Mukhtar, S.M. and Dawson, P.J. (1989). Herd size and unit costs of production in the England and Wales dairy sector, *Journal of Agricultural Economics*, 41, 9-20.

Shumway, R., Pope, R. and Nash E. (1984). Allocatable fixed inputs and jointness in agricultural production: implications for economic modelling, *American Journal of Agricultural Economics*, 66, 72-78.

Tyler, G. (1966). The use of multiple regression analysis of whole farm data in the estimation of enterprise labour coefficients, The Farm Economist, 11, 106-119.

Chapter eight: Future Directions for Multi-Sectoral Modelling and Rural Economics

Peter Midmore
Welsh Institute of Rural Studies, The University of Wales, Aberystwyth

Introduction

Contemporary economic forecasting methods are being criticised as a result of an increasing perception of their poor performance, mathematical complexity and even irrelevance, in the face of the political and economic problems that face economies in general.[1] These problems are mirrored when the particular case of the rural economy is considered. Therefore it might seem imprudent to offer a hostage to fortune by attempting to forecast the development of forecasting methods themselves. Nevertheless, there are important reasons why this can and should be attempted. Attempts by modellers to analyse and predict are driven, primarily, by problems that perplex policymakers, and it is not difficult to make out a convincing case of what those might be, in the rural sphere, in the medium term.

For instance, there are both external and internal pressures on European rural – chiefly agricultural – policy that have already initiated profound shifts in its character. There is a danger, of course, that the rôle of agriculture as a pivotal link in the production chain may be neglected. The 1993 GATT settlement requires considerable decoupling and reduction of agricultural support that will have major consequences for rural economies even though the farming sector may itself be partially insulated from them. Within the European Union itself, current arrangements benefit the northern members most, yet it is the southern members that have more serious structural problems in their extensive rural areas. These two forces suggest that reliance will shift to direct support measures, regional policy instruments and environmental incentives to resolve this complex of interdependent difficulties. Traditionally, there has been a focus on

1 See, for example, Ormerod (1994) especially chapter 5, pp. 92-112.

agriculture, rather than the rural economy as a whole. However, a number of contributions to this book have anticipated a shift to a wider perspective, examining alternatives such as rural tourism, information technology and the impact of enhanced infrastructures. Even so, further potential exists for analytical methods that allow comparison between the effectiveness of different kinds of intervention to achieve rural policy objectives.

Other policy issues relate to the environment of rural areas. To some extent these are linked to the grand question of trade liberalisation and how it affects rural economies: at least opportunistically, because they open up 'green boxes'. Extension of multi-sectoral economic modelling to include natural environmental systems has so far been explored inadequately. Equally, though, much of the current discussion of policy development is dominated by the difficulty of applying the principle of sustainability. By refocusing its capability for structural analysis, the multi-sectoral modelling approach has much to offer here.

Methodological issues also need to be confronted. With increasing computing power, there will be the temptation to construct ever more complex models and analyses, despite the problems of availability and reliability of rural economic data. Achieving the right balance between accuracy, ease of operation and credibility is likely to require considerable skill and judgement in the future.

It should be clear that the discussion of these issues is a blatant attempt to influence thinking about how modelling techniques *ought* to develop, rather than a dispassionate evaluation; it provides an agenda for inquiry by a new generation of investigators, as well as for the policymakers (and others who seek to influence the process of policymaking) who sponsor and initiate research. As a consequence, this chapter, being speculative and predominantly subjective, is qualitatively different in character from the material in the rest of the book. Hopefully, it will nevertheless provide an appropriate discussion and conclusion to it. The chapter is divided into three main sections: the first discusses modelling approaches that are appropriate for the analysis of policy-induced diversification of rural economies; the second reviews attempts to integrate input-output models with the environment, defined in conventional terms, and suggests ways in which such exploratory attempts might be developed; and the third examines the potential contribution that multi-sectoral methods of analysis might make to the assessment of rural economic sustainability.

Appropriate methods for the analysis of economic diversification

Rural economies are characterised by both peripherality and dependence on a narrow range of industries. Consequently, their susceptibility to economic leakages makes them far more vulnerable than would otherwise be the case. Taking both of these factors together, the resulting losses of people, predominantly in younger age groups, provide elements of an account of the cycle of cumulative causation that provides some explanation for the (seemingly intractable) socio-economic problems of the countryside. Attempts to reverse that cycle are costly, and often the close examination needed to justify expenditure includes an analysis of indirect effects. This section examines the appropriate mode of analysis for economic linkage effects in rural areas, and argues that these are particularly dependent on the scale of the economy being examined.

The smaller the scale of the economy, the narrower the local productive base is likely to be and therefore the lower the degree of linkage in production activities. At such a level, use of input-output techniques is probably inappropriate, because the overwhelming proportion of impact will be on production activities outside the boundary of the local economy in question. This narrowness of the economic base may lead to further problems since it is both difficult and costly to accurately measure the extent of non-local use of inputs in local production activities; and even if it were possible to do so, over the course of time non-local input usage would be unstable. In such circumstances, due to the relationship between the magnitude of impacts and the cost of measuring them, it is probably appropriate to focus on the impacts of additional consumption expenditures, which are much more likely to generate local indirect effects. The neo-Keynesian approach to calculation of local income and employment multipliers has been refined to a considerable degree, and a generalised approach that takes into account all likely variations in the local state of affairs has been developed by Ashcroft *et al.* (1988).[2]

At a regional scale, where inter-industry linkages may be created or enhanced by new activity (or in the opposite case, where declining activity may transmit reductions in production and incomes) use of the conventional input-output method is probably worthwhile. The high cost of estimating input-output coefficients in regions for which a flow matrix does not exist is an obvious problem that needs to be overcome. The methodical combination of use of survey methods for the most important coefficients, and mechanical coefficient reduction for the rest, reviewed by Jensen (1990), has had considerable recent appeal. This GRIT (Generating Regional Input-output Tables) framework has been somewhat abused by neglect of the need to use superior information[3] and it is necessary to stress the difficulties that arise when purely mechanical coefficient reduction techniques are used (for a review of the more important objections, see Round, 1983; however, modified procedures which produce improved results are discussed by Flegg *et al.*, 1995).

Within a small, rural region the standard input-output model, with its emphasis on linearity, absence of scale economies and aggregation of institutional flows is probably an appropriate compromise: however, where the analysis focuses on key distributional questions, or where the region under

---

2 For a comprehensive review of the general literature relating to measurement of economic impact, see Lewis (1988).

3 Though there are some notable exceptions in the analysis of rural policy changes: see, for example, Leat and Chalmers (1991).

consideration is large compared to the economy of which it forms a part, aspects of generality that might be reflected in a computerised general equilibrium or social accounts-based model increase in importance (for example, increasingly one might consider the UK economy as a region of the European Union for these purposes).

In general, the sophistication of modelling approach might be thought of as being increasingly justified in proportion to the scale and diversity of the economy in question:

- at a local level, often the degree of economic integration is quite limited, and as a result linkages related to production flows are likely to be negligible; in such circumstances, the use of neo-Keynesian local income and employment multipliers is most appropriate;

- at a sub-regional or regional level, linkages in production will increase in relation to consumption feedbacks, and when the economic importance of the area is small, relative to national aggregates, the restrictive Leontief technology assumptions are less serious in their implications; there are problems with incorporating consumption multipliers into an input-output framework, and whilst these can be offset by the SAM extension the cost is an increased data requirement;

- at the level of large regions, or even states within multinational trading blocs, CGEs may be more useful in coherently modelling the impact of major shifts in competitiveness. The lack of sectoral detail means that only very broad estimates of overall impact can be estimated.

Rural economies, by their nature, are likely to be at the lower end of the scale and thus more appropriately analysed in the neo-Keynesian or simple input-output framework: exceptionally, a SAM-based approach may be appropriate. However, note that in the case of *sectoral* analyses, as for agriculture, fishing and other industries based in rural areas, the predisposition to use less sophisticated methods does not necessarily apply.

Input-output and the rural environment

Substantial effort has been made in the past towards an integration of pollution and natural resources utilisation within an input-output framework. This included ambitious conceptual schemes that attempted to extend the social accounting concept to encompass economy-environmental system interaction, as represented in Figure 8.1.

	economic outputs	*pollutant outputs*
economic inputs	Production × Production	Production × Pollutants
natural resource inputs	Natural Resources × Production	Ecology × Ecology

Fig. 8.1. An economy-environment input-output matrix.

Several of the proposed variants of this approach have been summarised by Briassoulis (1986). The major obstruction to progress, in most cases, is the difficulty of adequately modelling the flows described in the bottom right quadrant of Figure 8.1, since they are predominantly non-linear, dynamic ecological systems that cannot be satisfactorily predicted using fixed, linear input-output coefficients. Without describing these, however, the model cannot be closed: attempts to link production activities with the environment have typically fallen back to models resembling Victor's (1972) exploratory extension of the input-output model as a process. This dealt with natural resource inputs as an extension to the primary inputs matrix, and described use of water resources as a waste sink separately, extending the final outputs matrix. These two interactions with the environment, however, were not explicitly linked. Recent examples of this approach, which treat pollution separately from resource extraction, may be found in Kulshreshtha's (1988) analysis of the contribution of the extraction of potash to the Saskatchewan economy, and Hawdon and Pearson's (1995) examination of the pollution output from energy industries in the United Kingdom. To a considerable extent, attempts to adjust national accounting procedures to include environmental capital depreciation (see United Nations Statistical Office, 1993[4]) could be enhanced by incorporation into an input-output framework.

The foregoing discussion leads, through the construction of satellite accounts, to the direct inclusion of environmental effect on welfare into income accounting procedures. This is controversial: a less ambitious, though relatively under stressed approach is to examine the consequences of making defensive

4 An example of such an approach for the agriculture and forestry industries has been undertaken in the United Kingdom by Adger and Whitby (1991).

expenditures on economic structure. Recent work on defensive expenditures for environmental protection by Nestor and Pasurka (1995) follows the approach of Leontief (1970) in describing anti-pollution economic sectors, using their inputs to link the cost of abatement to derived demands through the rest of the economy. Leontief's original system, set out in Table 8.1, incorporates estimates of the physical output of pollution and the industrial activity required to abate it. This internalisation of the production of pollution within the system allows it to be seen for what it actually is – part of the production process.

Table 8.1. Leontief's Pollution Abatement Model.

Inputs and Pollutant Output	Output Sectors Sector 1 Agriculture	Sector 2 Manufacturing	Anti-Pollution	Final H/hold Demand	Totals
Sector 1 Agriculture (bushels)	26.12 ($52.24)	23.37 ($46.74)	0.00	55.00 ($110.00)	104.50 ($208.99)
Sector 2 Manufacturing (yards)	14.63 ($73.15)	7.01 ($35.05)	6.79 (33.94)	30.00 ($150.00)	58.43 (292.13)
Pollutant (grams)	52.25	11.68	-33.93	30.00 ($101.80 paid for elimination of 33.93 grams of pollutant)	
Labour (man-years)	83.60 ($83.60)	210.34 ($210.34)	67.86 ($67.86)	0.00	361.80 ($361.80)
Column Totals	$208.99	$292.13	$101.80	$361.80	

Source: Leontief (1970: 268).

The table is couched primarily in physical terms in order to demonstrate the underlying productive relationships, especially as they apply to the production of pollutant. The solutions for the models based on these physical relationships allow an analysis of the technical interdependence between output and pollution, (in the inter-regional and dynamic versions of the model these can be extended over space and time respectively): this allows estimates of the direct and indirect abatement costs.

In the context of studies of the rural economy, where public good provision is increasingly obtained by direct purchases on the part of the state from farmers, the implications of this shift might well be analysed within the context of a similar framework. Most work that extends the insights of Leontief's approach have been concerned with emissions of pollutants.[5] Since such activities are interdependent with conventional agricultural production, and the construction of an 'environment-enhancement' sector, together with corresponding estimates of the output of public goods that result from it. An attempt to incorporate conservation activities into an input-output framework, made in the course of a wider analysis of pluriactivity (Bateman *et al.*, 1993), failed since none of the sampled farms had remunerated conservation activity on a significant scale. However, further development of the agri-environment measures introduced by the MacSharry reforms of 1992, together with regional schemes such as Countryside Stewardship are likely to have significantly increased the database necessary for estimating the general economic significance of this response to environmental concerns. The impact that conservation activities have on conservation structure might be included in the flow matrix in a similar manner by setting up a sector that, though consuming inputs, delivers zero output to final demand; the physical output of conservation benefits achieved can be included in the manner described in the mathematical appendix to Leontief's original article. Corresponding changes in relative prices and outputs can be calculated to reflect the opportunity cost of conservation.

Measuring rural economic sustainability

Since the Rio Summit in 1992, sustainability has become a ubiquitous if not platitudinous concept: its ambiguity has allowed it to become acceptable to widely differing shades of opinion, and it has taken on fresh nuances in the discourse of macro-economic policy where 'sustainable growth' refers not to resource husbandry but to economic growth without concomitant increases in price levels. Despite this catalogue of complaints, however, the commitment to a consideration of issues related to future resource availability is being taken seriously in the whole arena of policymaking. In this final section, the potential of the input-output approach in the analysis of the implications of the sustainability criterion is examined.

Lowe and Murdoch (1993) have considered some of the specific issues raised by sustainability for rural economies. Their discussion centred on two major themes: the changes necessary in rural resource utilisation; and the (simultaneous) changes in attitude required of developers, environmentalists and policymakers to make these changes effectively. In addition to traditional rural natural resource products (food, fibre, timber and minerals) the countryside is now increasingly expected to absorb urban wastes (exporting assimilative capacity) and meet intensifying demands for conservation and space for leisure

---

5 See, for example, the work of Nestor and Pasurka cited above: for other recent examples see Luptacik and Bohm, 1994, or Dufournaud *et al.*, 1988; an exception is provided by Gould, 1986, who analyses resource use in a similar context.

activities. The high inputs of energy (including that used to produce the hydrocarbon-based agrochemicals in agriculture) which are responsible for current environmental problems in the countryside will need to be reduced and shift towards renewable sources. The attitude changes required of developers involve increased environmental awareness, through energy audits and environmental responsibility, which is increasingly sought by major businesses throughout the product life cycle in which they are involved. Environmentalists need to recognise the reciprocity between developments that protect and improve employment opportunities and more sustainable rural environmental systems. Policymakers will be required to take the interconnections between transport, settlement, employment and demography more and more into account, especially as the rising real costs of private motor transport fall ever more on rural people, and to promote a shift of employment into the countryside and improve the stock of dwellings available to younger people, built to the highest energy-saving standards.

These arguments are familiar, and made more effective in the rural context as the symptoms of unsustainability are clearly recognisable. In addition to hazardous dependence on non-renewable resources and degradation and decline in the availability of renewable resources, there is also depopulation (or at least increasing demographic bias towards the elderly), and economic dependence on continuing and rising levels of subsidy from an urban population which is increasingly disinclined to pay up. Crucially, however, the analysis lacks insights into the *causes* of these symptoms. Though a full discussion of the origin and nature of rural economic decline is beyond the scope of this section, some indication of the themes involved is necessary to show how input-output analysis could be used to support decision-making that contributes to reversing the trend.

Spatially uneven development is often attributed to cycles of cumulative causation (Myrdal, 1957) based initially on coincidental or accidental combinations of resource availability. Rural areas tend to have the wrong kind of economic diversity. Consequently, (although there are notable exceptions) they suffer from backwash effects, with low levels of investment reducing employment opportunities, lowered expectations and hopelessness. With increasing concentrations of urban wealth, trade liberalisation and developments in the transport of bulky goods over long distances, the conditions of exchange between urban and rural areas are affected, most often to the detriment of the latter. In the context of nation states, conditions such as these have given rise to arguments for protection: though trade barriers are obviously infeasible, the means to promote self-reliance and resilience for rural economies could be had by deliberately cultivating comparative advantage, and by seeking to create what are effectively non-tariff barriers. Comparative advantage can be sought through the encouragement of locally distinctive products incorporating relatively high levels of high quality labour input. Green consumerism is the obvious counterpart to such activities, but it has limitations; in particular it is fashionable and prone to decline during recessionary periods. Because to a large extent, employment prospects in the countryside depend on small- to medium-scale activities, and (as discussed above) rural economies tend to suffer greater levels of leakage than their urban counterparts, a means of offsetting peripheral rural economic vulnerability would be to deliberately create linkages by fostering mutual intra-rural interdependence between activities. Because input-output analysis is a tool for measuring and evaluating linkages, its framework can be adapted to maximise employment-generating activities at the local rural level (see, for example, Sonis and Hewings, 1993). Of course, it is essential that concerns related to accuracy discussed above and in Chapter two are taken into account, and technical developments in the availability of information and the estimation of inflows and outflows of goods and services are required if maximum effectiveness is to be captured.

Conclusion

Multi-sectoral modelling within the tradition established by Leontief is a flexible tool: opportunities for its use in a rural context are widespread, and it can be adapted to assist in the resolution of a number of urgently pressing problems. Little further by way of conclusion is needed, since this reveals a substantial scope for further extension and development. In this volume we have set out to discuss most of the relevant issues, in particular arguing that the weaknesses of the approach should not be neglected by attempting applications in inappropriate contexts. Nevertheless, other approaches have drawbacks that are even more serious; the contributions here, which have analysed the diverse range of analysis possible, demonstrate a commitment to a firmly realistic applied economics.

References

Adger, N. and Whitby, M. (1991). Accounting for the impact of agriculture and forestry on environmental quality, *European Economic Review*, 35, 629-641.

Ashcroft, B., Holden, D., Smith, J. and Swales, J.K. (1988). A generalised augmented keynesian local multiplier model. In: Harrigan, F. and McGregor, P.G. (eds.) *Recent advances in regional economic modelling*, London: Pion, 14-23.

Bateman, D.I., Hughes, G.O., Midmore, P., Lampkin, N.H. and Ray, C. (1993). *Pluriactivity and the rural economy in the less favoured areas of Wales*, Report to the ESRC, Aberystwyth: Department of Economics and Agricultural Economics (University of Wales).

Briassoulis, H. (1986). Integrated economic-environmental-policy modeling at the regional and multiregional level – methodological characteristics and issues, *Growth and Change*, 17, 22-34.

Dufournaud, C.M., Harrington, J.J. and Rogers, P.P. (1988). Leontief environmental repercussions and the economic-structure revisited – a general equilibrium formulation, *Geographical Analysis*, 20, 318-327.

Flegg, A.T., Webber, C.D. and Elliott, M.V. (1995). On the appropriate use of location quotients in generating regional input-output tables, *Regional Studies*, 29, 547-561.

Gould, B.W. (1986). The impacts of structural-change within an economy on resource use – an input-output-analysis, *Applied Economics*, 18, 457-477.

Hawdon, D. and Pearson, P. (1995). Input-output simulations of energy, environment, economy interactions in the UK, *Energy Economics*, 17, 73-86.

Jensen, R.C. (1990). Construction and use of regional input-output models – progress and prospects, *International Regional Science Review*, 13, 9-25.

Kulshreshtha, S.N. (1988). Estimation of contributions of a resource sector to a provincial economy – the case of Saskatchewan potash, *Canadian Journal of Regional Science-Revue Canadienne des Sciences Regionales*, 11, 431-444.

Leat, P.M.K. and Chalmers, N. (1991). Analysing agricultural adjustment in Grampian using an input-output model of the agricultural and food complex. In: Midmore, P. (ed.), *Input-output models in the agricultural sector*, Aldershot: Avebury, 57-71.

Leontief, W.W. (1970). Environmental repercussions and economic structure: an input-output approach, *Review of Economics and Statistics*, 52, 262-271.

Lewis, J.A. (1988). Economic-impact analysis – a UK literature survey and bibliography, *Progress in Planning*, 30(P3), 159-209.

Lowe, P. and Murdoch, J. (1993). *Rural sustainable development: report for the Rural Development Commission*, London: Rural Development Commission.

Luptacik, M. and Bohm, B. (1994). An environmental input-output model with multiple criteria, *Annals of Operations Research*, 54, 119-127.

Myrdal, G (1957). *Economic theory and under-developed regions*, London: Duckworth.

Nestor, D.V. and Pasurka, C.A. (1995). Environment economic accounting and indicators of the economic importance of environmental-protection activities, *Review of Income and Wealth*, 3, 265-287.

Ormerod, P. (1994) *The death of economics*, London: Faber and Faber.

Round J.I. (1983). Nonsurvey techniques – a critical review of the theory and the evidence, *International Regional Science Review*, 8, 189-212.

Sonis, M. and Hewings, G.J.D. (1993). Hierarchies of regional sub-structures and their multipliers within input-output systems – Miyazawa revisited, *Hitotsubashi Journal of Economics*, 34, 33-44.

United Nations Statistical Office (1993). *Integrated environmental and economic accounting*, Studies in methods Series F, New York: United Nations.

Victor, P.A. (1972). *Economics of pollution*, London: Macmillan.

Index